FREE DVD  FREE DVD

From Stress to Success DVD from Trivium Test Prep

Dear Customer,

Thank you for purchasing from Trivium Test Prep! Whether you're a new teacher or looking to advance your career, we're honored to be a part of your journey.

To show our appreciation (and to help you relieve a little of that test-prep stress), we're offering a **FREE** ***AFOQT Essential Test Tips DVD*** * by Trivium Test Prep. Our DVD includes 35 test preparation strategies that will help keep you calm and collected before and during your big exam. All we ask is that you email us your feedback and describe your experience with our product. Amazing, awful, or just so-so: we want to hear what you have to say!

To receive your **FREE** ***AFOQT Essential Test Tips DVD***, please email us at 5star@triviumtestprep.com. Include "Free 5 Star" in the subject line and the following information in your email:

1. The title of the product you purchased.
2. Your rating from 1 – 5 (with 5 being the best).
3. Your feedback about the product, including how our materials helped you meet your goals and ways in which we can improve our products.
4. Your full name and shipping address so we can send your **FREE** ***AFOQT Essential Test Tips DVD***.

If you have any questions or concerns please feel free to contact us directly at 5star@triviumtestprep.com.

Thank you, and good luck with your studies!

* Please note that the free DVD is not included with this book. To receive the free DVD, please follow the instructions above.

AFOQT Study Guide 2018-2019

2018-2019

AFOQT Exam Prep and Practice Test Questions for the Air Force Officer Qualifying Test

TABLE OF CONTENTS

INTRODUCTION

Congratulations on preparing to take the first step to becoming an officer in the United States Air Force! The AFOQT is an important exam for many reasons, but mostly because it lets the Air Force know where your strengths and weaknesses lie, and how they can best utilize your abilities. It gives them a glimpse of your raw abilities in a few subjects, but will also test your ability to stay calm under pressure.

Some sections, such as Math and Science, will test your raw knowledge and intellect, not to mention how hard you prepared (or not) for the exam. Other sections are deceptively simple, in that the task is something you might think a six-year-old could do. The catch though is that you have an absurdly short amount of time on these sections and the task is usually designed to seem easier than it is. Their goal, in this case, is to see how you respond when the odds are stacked against you, and you know that you won't be able to finish. Do you panic and make mistakes, or push through diligently and do your best?

This study guide will prepare you for both aspects of this challenging exam. The AFOQT is a competitive test, so you need to aim high… no pun intended.

Sections on the AFOQT

There are twelve sections on the AFOQT and will follow this order:

1. Verbal Analogies: 25 questions, 8 minutes
2. Arithmetic Reasoning: 25 questions, 29 minutes
3. Word Knowledge: 25 questions, 5 minutes
4. Math Knowledge: 25 questions, 22 minutes
5. Instrument Comprehension: 20 questions, 6 minutes
6. Block Counting: 20 questions, 3 minutes
   Ten-minute break
7. Table Reading: 40 questions, 7 minutes
8. Aviation Information: 20 questions, 8 minutes
9. General Science: 20 questions, 10 minutes
10. Rotated Blocks: 15 questions, 13 minutes

11. Hidden Figures: 15 questions, 8 minutes

12. Self-Description Inventory: 220 questions, 40 minutes

As you can see, there are a few sections with minimal time limits, giving you only ten to fifteen seconds per question in some cases. The only section we will NOT cover in this study guide is *Self-Description Inventory*. This is not a section you can study for and there are no tricks here. It is simply a personality questionnaire, so be sure to be totally honest and straightforward in your answers. Don't give answers for what you think they want to hear.

Scoring on the AFOQT

You will be given scores in five areas:

- Pilot
- Navigator
- Academic Aptitude
- Verbal
- Quantitative

Each score is made up of different combinations of the twelve sections of the AFOQT and each has a different minimum score. In all five cases, the scores are percentiles, which range from 1 – 99. An average score would fall in the upper 40s. It is recommended that for those who want to be a pilot, that you achieve a score of 70 or higher across all five categories. Scores in the 85 – 95 range are considered strong scores. The sections comprising each score and the minimum required score to be considered are listed below for your reference:

Pilot

The Pilot score is comprised of five sections, including Artithmetic Reasoning, Math Knowledge, Instrument Comprehension, Table Reading, and Aviation Information.

- minimum score to be Pilot: 25
- minimum combined Pilot and Navigator score: 50 (must have at least 10 Navigator score)

Navigator

The Navigator score is comprised of six sections, including Verbal Analogies, Arithmetic Reasoning, Math Knowledge, Block Counting, Table Reading, and General Science.

- minimum score to be Navigator: 25
- minimum combined Pilot and Navigator score: 50 (must have at least 10 Pilot score)

Academic Aptitude

The Academic Aptitude score is comprised of four sections, including Verbal Analogies, Arithmetic Reasoning, Word Knowledge, and Math Knowledge. There is no minumum score; it is a composite of Math and Verbal sections.

Verbal

The Verbal score is comprised of two sections, including Verbal Analogies and Word Knowledge. The minimum score for all candidates is 15.

Quantitative

The Quantitative score is comprised of two sections, including Arithmetic Reasoning and Math Knowledge. The minimum score for all candidates is 10.

Test Day Procedures

On test day, be prepared for a three-and-a-half hour test day, including administrative information and instructions. You will have one ten minute break halfway through the exam. You do not want to waste time reading instructions; you should be totally familiar with those before sitting down. Pencils and paper will be provided; you may NOT use a calculator. No food or drink is allowed during the test.

It is important to note that there is no wrong answer penalty on the AFOQT. Again, **THERE IS NO PENALTY FOR WRONG ANSWERS**! This means, under no circumstance, on any section of the entire test, should you EVER leave a question blank even if it is a wild guess.

Additional Important Information

One of the most important factors to take into consideration is that you may only take the AFOQT twice. Never, ever think of the first exam as a "practice round." You need to study diligently and do your absolute best the very first time you take it.

If you decide to take the exam a second time, keep in mind you must wait 180 days (six months), and your second score will become your official score, even if you do worse! Your most recent score is the only one anyone will see. In some extremely rare circumstances, you might secure a waiver to take the AFOQT a third time, but that is highly unlikely for most candidates.

Finally, it is important to consider that the USAF looks at the whole package you send them. For example, if the average score for candidates who earn a pilot position is 90, just because you might only have an 85 does not mean it is impossible to win a spot. Conversely, just because you have a 99 does not automatically guarantee anything. They will look at your entire application package, including your GPA and the other four scores you receive.

One more time, just to be sure: DO NOT EVER LEAVE A QUESTION BLANK! You will better understand as you work through the different sections just how short the time limits are. Do not allow yourself to be caught off guard with only five seconds left and not enough time to fill in any answers before the buzzer. You must maintain situational awareness of the clock, which is best done by practicing under timed conditions.

About Accepted, Inc.

Accepted, Inc. uses industry professionals with decades' worth of knowledge in their fields, proven with degrees and honors in law, medicine, business, education, the military, and more, to produce high-quality test prep books for students.

Our study guides are specifically designed to increase any student's score, regardless of his or her current skill level. Our books are also shorter and more concise than typical study guides, so you can increase your score while significantly decreasing your study time.

How to Use This Guide

This guide is not meant to waste your time on superfluous information or concepts you've already learned. Instead, this guide will help you master the most important test topics and also develop critical test-taking skills. To support this effort, the guide provides:

- organized concepts with detailed explanations
- practice questions with worked-through solutions
- key test-taking strategies
- simulated one-on-one tutor experience
- tips, tricks, and test secrets

Because we have eliminated the filler and fluff, you'll be able to work through the guide at a significantly faster pace than you would with other test prep books. By allowing you to focus only on those concepts that will increase your score, we'll make your study time shorter and more effective.

We Want to Hear from You

Here at Accepted, Inc. our hope is that we not only taught you the relevant information needed to pass the exam, but that we helped you exceed all previous expectations. Our goal is to keep our guides concise, show you a few test tricks along the way, and ultimately help you succeed in your goals.

On that note, we are always interested in your feedback. To let us know if we've truly prepared you for the exam, please email us at support@acceptedinc.com. Feel free to include your test score!

Your success is our success. Good luck on the exam and your future ventures.

Sincerely,

– The Accepted, Inc. Team –

one

VERBAL ANALOGIES

Some relationship objectives are contained in the Verbal Analogies section. With these types of analogies, you must determine the relationship between the words and complete the analogy with the correct choice.

There are twenty-five questions with a time limit of eight minutes, giving you roughly twenty seconds per question. The key here is to become familiar with analogies and get some practice, and most people will find that twenty seconds is ample. We will first review some of the types of analogies you might run into, and then do a practice test. The concept of analogies is pretty straight forward, but don't brush it off because it is the subtleties that lose points.

Similarity and Contrast

Most often, similarities and contrasts involve synonyms (words that mean the same thing) or antonyms (words that have opposite meanings). An analogy that involves synonyms is primarily a definition of terms—determining one word that could be replaced with another word. In such a relationship, you must ascertain what a word means and how it is connected to the others in the analogy. An analogy that involves contrasts shows the relationship between a word and its opposite.

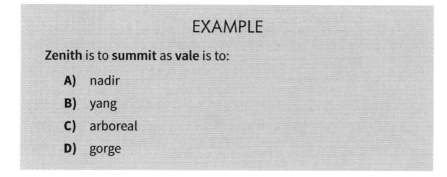

EXAMPLE

Zenith is to **summit** as **vale** is to:

- **A)** nadir
- **B)** yang
- **C)** arboreal
- **D)** gorge

Whole-Part and Part-Whole

This type of analogy denotes the relationship between a whole thing (e.g., house) and a part of the whole (e.g., room).

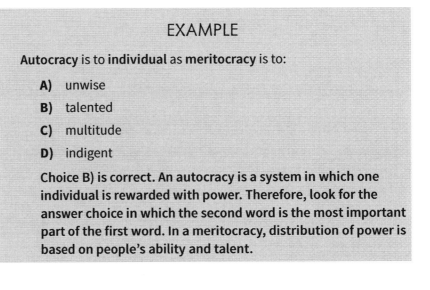

EXAMPLE

Autocracy is to **individual** as **meritocracy** is to:

A) unwise

B) talented

C) multitude

D) indigent

Choice B) is correct. An autocracy is a system in which one individual is rewarded with power. Therefore, look for the answer choice in which the second word is the most important part of the first word. In a meritocracy, distribution of power is based on people's ability and talent.

Membership

A membership analogy is very similar to the whole-part analogy. It shows the relationship between a whole group and a member of the group.

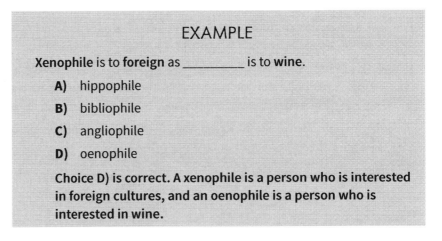

EXAMPLE

Xenophile is to **foreign** as _____ is to **wine.**

A) hippophile

B) bibliophile

C) angliophile

D) oenophile

Choice D) is correct. A xenophile is a person who is interested in foreign cultures, and an oenophile is a person who is interested in wine.

Object and Characteristic

In this type of analogy, you must establish the relationship between a person or object and its characteristic.

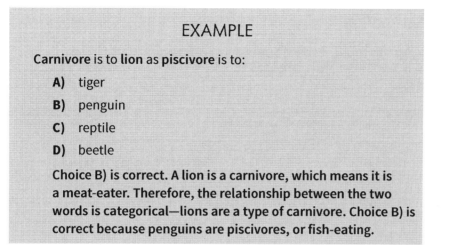

EXAMPLE

Carnivore is to **lion** as **piscivore** is to:

A) tiger

B) penguin

C) reptile

D) beetle

Choice B) is correct. A lion is a carnivore, which means it is a meat-eater. Therefore, the relationship between the two words is categorical—lions are a type of carnivore. Choice B) is correct because penguins are piscivores, or fish-eating.

Cause and Effect

This type of analogy involves analyzing the relationship between a word and the outcome or result it causes. Occasionally, the analogy may be written with the effect first, and you must determine the cause.

EXAMPLE

Sycophant is to **flatters** as **raconteur** is to:

A) critiques

B) repels

C) regales

D) leads

Choice C) is correct. In the initial part of the analogy, a sycophant is a person who flatters others. So, in the answer choices, look for the pair in which the second word describes the effect of the first term's behavior. A raconteur is a storyteller—one who regales others with tales.

Agent and Object

This analogy type is one that shows the relationship between a person and a tool or object that he/she uses. You might also see similar analogies that involve a non-living thing—an object—and how it is used.

GO ON

Order

In order analogies, the words are related by sequence or in a reciprocal (or opposite) circumstance.

Practice Problems

1. **Carat** is to **weight** as **fathom** is to:

    1-A   capacity

    1-B   perspective

    1-C   mass

    1-D   depth

2. **Mollify** is to **enrage** as **quell** is to:

    2-A   exonerate

    2-B   bifurcate

    2-C   incite

    2-D   denounce

3. **Electron** is to **satellite** as **planet** is to:

    3-A   proton

    3-B   nucleus

    3-C   atom

    3-D   neutron

4. **Habitat** is to **location** as _____ is to **role**.

    4-A   capacity

    4-B   competition

    4-C   niche

    4-D   predation

5. **26** is to **even** as _____ is to **prime**.

    5-A   8

    5-B   11

    5-C   15

    5-D   20

6. **Cytology** is to **cells** as _____ is to **fungi**.

    6-A   mycology

    6-B   ornithology

    6-C   oncology

    6-D   phrenology

7. **Darwin** is to **evolution** as **Mendel** is to:

    7-A   blood groups

    7-B   popular culture

    7-C   relative dating

    7-D   genetics

8. **Sacramento** is to **California** as _____ is to **Florida**.

    8-A   Orlando

    8-B   Tallahassee

    8-C   Miami

    8-D   Jacksonville

9. **Appease** is to **placate** as **obviate** is to:

    9-A   disregard

    9-B   clarify

    9-C   decide

    9-D   preclude

10. **Cortez** is to **Mexico** as **Cartier** is to:

    10-A   North America

    10-B   Canada

    10-C   East Indies

    10-D   Florida

11. **Chicken** is to **brood** as **cats** are to:

    11-A   clowder

    11-B   herd

    11-C   swarm

    11-D   army

12. _____ is to **Missouri** as **Granite** is to **New Hampshire**.

    12-A   Tar Heel

    12-B   Sunshine

    12-C   Show Me

    12-D   Buckeye

GO ON

13. **Latitude** is to **longitude** as **parallels** is to:

  13-A   lines

  13-B   equator

  13-C   degrees

  13-D   meridians

14. *Gulliver's Travels* is to **Jonathan Swift** as *Frankenstein* is to:

  14-A   John Keats

  14-B   George Eliot

  14-C   Mary Shelley

  14-D   Alexander Pope

15. **Newton** is to **force** as _____ is to **power**.

  15-A   joule

  15-B   tesla

  15-C   watt

  15-D   hertz

16. **0° latitude** is to **equator** as **23.5° south latitude** is to:

  16-A   prime meridian

  16-B   tropic of Capricorn

  16-C   middle latitude

  16-D   tropic of Cancer

17. **Clarinet** is to _____ as **cello** is to string.

  17-A   brass

  17-B   woodwind

  17-C   percussion

  17-D   horn

18. **Hatching** is to **parallel lines** as **stippling** is to:

  18-A   squares

  18-B   dots

  18-C   perpendicular lines

  18-D   curves

19. **Capacity** is to **liter** as **mass** is to:

  19-A   meter

  19-B   kilometer

  19-C   gram

  19-D   pound

20. **Inaugurate** is to **President** as **coronate** is to:

  20-A   pope

  20-B   cardinal

  20-C   bishop

  20-D   monarch

21. **Positive** is to **negative** as _____ is to **flat**.

  21-A   sharp

  21-B   tone

  21-C   bass

  21-D   treble

22. **Poseidon** is to **sea** as _____ is to **war**.

  22-A   Hades

  22-B   Ares

  22-C   Apollo

  22-D   Pan

23. **Hindi** is to **India** as _____ is to **Brazil**.

  23-A   Spanish

  23-B   English

  23-C   Portuguese

  23-D   French

24. **Michelangelo** is to **painter** as **Frank Lloyd Wright** is to:

  24-A   writer

  24-B   sculptor

  24-C   playwright

  24-D   architect

25. **Judaism** is to **temple** as **Islam** is to:

25-A   church

25-B   synagogue

25-C   mosque

25-D   mecca

Verbal Analogies
Answer Key

1.	D	14.	C
2.	C	15.	C
3.	B	16.	B
4.	C	17.	B
5.	B	18.	B
6.	A	19.	C
7.	D	20.	D
8.	B	21.	A
9.	D	22.	B
10.	B	23.	C
11.	A	24.	D
12.	C	25.	C
13.	D		

two

ARITHMETIC REASONING

Introduction

The **ARITHMETIC REASONING** subtest of the AFOQT is primarily comprised of mathematical word problems. The general purpose of this test is to determine how well you can apply your mathematical knowledge to situations you may encounter in the real world. It is important to note that the questions you encounter on this section of the test involve both reasoning (logic) and arithmetic. They are not as complicated as the ones that you might encounter in the other math portion of the AFOQT, but they may be complicated since the actual problem may not be directly stated.

This is one of the only test sections you are likely to encounter that will necessitate the need of the scratch paper that you have been provided. You will NOT be allowed to use a calculator, so you should practice these types of problems without the use of one to be prepared.

Qualification for many jobs within the military depends on your success on the arithmetic reasoning test as well. The following line scores utilize the arithmetic reasoning score:

Table 2.1. Line scores by military branch

BRANCH	LINE SCORE
Army	Clerical, general technical, skilled technical, operators and food, surveillance and communications
Marines	General technical
Navy/Coast Guard	Administrative, health, nuclear, general technical
Air Force	Administrative and general

There are three types of questions you are likely to encounter on the arithmetic reasoning section of the AFOQT: algebra word problems, fact-finding word problems, and geometry word problems.

ALGEBRA WORD PROBLEMS are word problems that necessitate the creation of an equation with an unknown in order to solve them. These are not too complex and, generally speaking, you will not have to use too much algebra to solve these.

EXAMPLE

John spent $42 on a pair of shoes. That was $14 less than twice the amount of money John spent on a new pair of Jeans. How much did the jeans cost?

As you can see, you are trying to find a value that isn't here. You would best set this problem up in the following way, the unknown being x:

$2x - 14 = 42$

In other words, 2 times x (the value of the jeans) minus 14 will yield **42**, the value of the shoes that John bought. Do you see how you can "translate" the word problem into math, almost word by word? And at this point, you'll probably want to plug in the answer choices to see which one works. **The correct answer is 28**, since 2 × 28 − 14 = 42. You can also solve this using more complicated algebra, as we'll see in the mathematics knowledge section.

The second type, FACT-FINDING WORD PROBLEMS, are problems that state facts and then ask you to do something with them.

EXAMPLE

The chess club is trying to raise money by selling chocolates that are shaped like chess pieces. Jimmy sold 3 of them, Sally sold 5, and John sold 12. How many chocolates were sold in total?

First, determine the facts:

Jimmy = 3 chocolates

Sally = 5 chocolates

John = 12 chocolates

Total chocolates sold = ?

Next, simply add them together.

3 + 5 + 12 = 20

20 chocolates were sold in total.

The third type, GEOMETRY WORD PROBLEMS, are problems asking you to find a perimeter or area of a given space. Typically, these will not be more complex than that.

With a firm grasp of these types of questions, you will be prepared for the arithmetic reasoning subtest questions.

Numbers

Whole Numbers

Most people know what whole numbers are, even if you do not know the term "whole number". A whole number is exactly what it sounds like: an *entire* number. Examples of whole numbers are 0, 1, 2, 3, 4, and so on. There is no decimal or fractional part. Whole numbers are part of a system that is known as the place value system. In this system, numbers are labeled with specific words, depending on their position.

You may have heard the term "counting numbers" or "natural numbers," which are whole numbers without the "0". Natural numbers are often used for basic counting, ordering, and things of that nature. These numbers are the simplest form of numbers, from which other number sets are created.

Here is a brief table you can use to refresh your memory and better grasp this concept:

Table 2.2. Whole numbers

VALUE	PLACE
one	1
ten	10
hundred	100
thousand	1,000
ten thousand	10,000
hundred thousand	100,000
million	1,000,000

This decimal number system (also called base 10 since it has 10 as its base) continues into the millions and beyond.

If you were given a number such as 14,567, the word associated with it would be fourteen thousand, five hundred and sixty-seven. It is important to grasp this concept because on the arithmetic reasoning portion of the AFOQT, you may encounter numbers as their long-form versions rather than the punctual versions you are used to.

ROUNDING is a term that is used to quickly determine the relative size of a whole number based on a new number (either larger or smaller than the original number). Rounding is the best method to approximate numbers quickly. Below is the general procedure for rounding:

- Choose the digit you will be rounding and underline it (so you know which one it is).
- Look at the digit directly to the right of the digit you are rounding. If that number is less than 5, then you will leave the underlined digit alone. If the digit to the right of the digit you are rounding is 5 or more, then you will add 1 to the digit you are rounding.
- Last, replace all of the digits that are to the right of the underlined digit with 0s (up to the decimal point).

EXAMPLE

Round 356,782 to the nearest ten thousandth.

First you will select the ten-thousands spot, the 5. Then you will look at the number to the right of it, the 6. That number is larger than 4, so you add one to the 5 spot, giving you 366,472. Now you will replace numbers to the right of the primary digit with 0's, giving you **360,000**.

Fractions and Decimals

Fractions and decimals are both methods of elaborating on *parts* of whole numbers. This is usually illustrated by two numbers, say 0 and 1, and then viewing them via a line between them. That line represents all of the fractions or decimals in between those two numbers.

Take FRACTIONS first. The most basic way to view a faction is to look at it as parts of a whole. The division sign ("/" or "−"), which separate the two numbers, can represent the words *out of*. So if you have $\frac{2}{4}$, you have 2 out of 4 parts, which is actually the same as 1 out of 2 parts. For example:

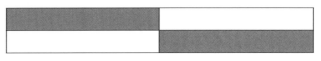

Figure 2.1. Two out of four parts

In the above figure, you can see that there are 4 parts (rectangular boxes). Of these, 2 of them are gray. So you have $\frac{2}{4}$ boxes that are gray.

Note that this also represents $\frac{1}{2}$ of the boxes, and we can say that $\frac{2}{4}$ reduces or simplifies to $\frac{1}{2}$.

The top part of the fraction is called the NUMERATOR and the bottom part is called the DENOMINATOR. When looking at fractions, if the numerator is less than the denominator, then it is referred to as a PROPER FRACTION. The value of these types of fractions is going to be less than 1. Some examples of proper fractions are $\frac{3}{9}$ or $\frac{1}{3}, \frac{2}{5}$, and $\frac{8}{19}$.

If the numerator is equal to or greater than the denominator, then you have an IMPROPER FRACTION. Some examples of improper fractions are $\frac{6}{3}$ or $\frac{2}{1}$ (which is 2), or $\frac{8}{3}$ and $\frac{5}{2}$. Any improper fraction can be written in the form of a MIXED NUMBER, and any mixed number can be written in the form of an improper fraction. For example, $\frac{4}{3}$ is the same as $\frac{3}{3} + \frac{1}{3} = 1\frac{1}{3}$.

It should be noted that with fractions, having $\frac{4}{4}$ is the same as having 1, $\frac{8}{4}$ is the same as 2, and $\frac{9}{3}$ is the same as 3; these pairs of fractions are EQUIVALENT. (See how if we were to multiply both the numerator and denominator by the same number, we will get equivalent fractions?) Likewise, having $\frac{6}{4}$ is the same as $\frac{3}{2}$ which equals $1\frac{1}{2}$. This is how fractions are converted back and forth into whole numbers and mixed fractions.

To review:

- Proper fractions: These are fractions with a value of less than 1. The numerator is smaller than the denominator. $\frac{1}{2}$ is an example.

- Improper fractions: These are fractions with a value of 1 or more. The numerator is larger than the denominator. $\frac{7}{5}$ is an example.

- Mixed numbers: These are whole numbers that are combined together with fractions. $5\frac{1}{2}$ is an example.

Now onto DECIMALS. Everything talked about earlier, in terms of whole numbers, have included digits that are powers of 10, but greater than 1 (to the left-hand side of the decimal point). Again, the number system is a BASE 10 SYSTEM. Decimals are a kind of shorthand that are used to describe base 10 numbers (fractions of a whole number) that are on the right-hand side of the decimal point, which means they are less than 1.

Here is a table which might help explain this concept a bit better. Notice that the names of the places to the right of the decimal include a *th* at the end:

Table 2.3. Decimals

PLACE	DECIMAL
tenth	0.1
hundredths	0.01
thousandth	0.001

In this way, you can describe numbers smaller than whole numbers in a method other than using fractions. This is also how percents are written (see the next section). So if you have 0.5, you have exactly $\frac{1}{2}$ of 1. You can also think of this as $\frac{5}{10}$, or $\frac{1}{2}$, or $\frac{.5}{1}$, which is five-tenths of 1.

Remember that typically in fractions and percent problems, *of* represents multiplication; for example, $\frac{1}{2}$ of 4 = $\frac{1}{2}$ × 4 = 2.

Percents

PERCENTS can be considered fractions that have a bottom number (denominator) equal to 100. You can think of this as *per hundred*, which is what *percent* actually means. The symbol that is used to denote a percent is %. For example:

$\frac{50}{100}$ is the same thing as .50, which is the same thing as 50%.

You can see that there is a very clear and definite relationship among fractions, decimals, and percents. Percents are typically used in order to describe portions of a whole, just like the other two are. If you have a fraction, you can convert the top and bottom numbers to $\frac{\#}{100}$ in order to determine the percent. Likewise, you can simply shift the decimal two places to the right to get a percent from a decimal; for example 0.75 is 75%. To get a decimal from a percent, shift the decimal two places to the left; for example, 45% is .45.

A typical question about percents on the AFOQT might be converting them to decimals and back to percents.

Operations

Operations are the calculations that are done to numbers. *The Fundamental Operations of Arithmetic*, as they are called, are addition, subtraction, multiplication, and division. Solving the problems in the AFOQT will require you to perform these operations on whole numbers, decimals, and fractions.

Addition and Subtraction

The ADDITION of whole numbers is not too complicated. Adding two numbers together is an operation that results in a total that is known as the sum. The very first step here is to line up the numbers you are going to add by their place value. All of the ones will go in a line (column), and then tens, hundreds, thousands, and so on. Once this has been done, you will add the columns together. Any time the sum of a column reaches 10, you need to carry over the 10 as a single digit to the next column to the left. So 10 in the ones column would be 1 in the tens column. Proceed to do this until the numbers have been added up.

EXAMPLE

Add together 20, 15, 2, and 107.

```
  20
  15
   2
+107
 144
```

Now look at those numbers. The digits in the ones column (the right-most column) add up to 14. Bring down the 4 to the ones digit and carry over the 10 to the next column (to the left). So 2 + 1 + 1 (the carryover) + 0 = 4. So there is a 4 in the tens digit as well. The hundreds column only has a 1. So the final result is **144**.

Adding together decimals is the exact same procedure. The only change is that you have to deal with that pesky decimal point. All you have to do is line up the decimal point vertically with every single number you need to add. You then add them together just like before to get your sum. For whole numbers, you can make this easier by adding a decimal point to the right of the number (so 17 would become 17.0 or 17.00, for example, depending on how long the other decimals are).

If you want to make things easier, you can just put 0s in the place of any empty spots to make it easier to add together.

EXAMPLE

Add together 13.2, 0.54, 2, and 0.008.

```
  13.2
   0.54
   2
+  0.008
```

The best way to do this is to modify the numbers a little bit. Note that they are still the same numbers, but easier to add:

```
  13.200
  00.540
  02.000
+ 00.008
  15.748
```

You can readily see that this small procedure makes these numbers much easier to add together. Adding them will give you **15.748.**

SUBTRACTION is the inverse or opposite of addition. It could also be considered another form of addition (by adding negative numbers), but that concept will be covered in a later section. Subtracting a number from another number will give you a result that is known as the difference. The steps are the same but backward. Line up the two numbers like before. (You should only subtract two numbers at a time). If the number on top is less than the one on the bottom for that column, then "borrow" from the next column (the one to the left) and continue with the operation.

EXAMPLE

Subtract 108 from 110.

$$110$$
$$-108$$
$$2$$

Start with the right-most column, the ones column. Since 8 is greater than 0, you have to borrow from the tens column to turn the 0 into a 10 before you subtract. But then you have to turn the 1 in the tens column into a 0 (subtract 1 in the tens column). So the final answer is just **2**. Again, like addition, this is a pretty simple process that most individuals are already used to doing without even thinking about it.

The subtraction of decimals, again, is much the same as the addition of decimals. Before subtracting, you have to line up the decimal points of the numbers that are being subtracted from one another and then place the decimal point below those two amounts in the same spot.

EXAMPLE

Subtract 89.3 from 109.4.

$$109.4$$
$$- 89.3$$
$$2$$

The first thing you want to do is fill in any zeroes to the right of the decimal point. In this case, we don't need to borrow.

$$109.4$$
$$- 089.3$$
$$\mathbf{020.1}$$

Practice makes perfect when it comes to addition and subtraction. Spend a little bit of time working on these types of problems on your own. The more you do it, the easier it will become. It would benefit you greatly to do these types of problems on paper without the use of a calculator. This will help you with mental calculations and will greatly increase your accuracy. There are a lot of websites that provide practice on these types of problems.

Multiplication

MULTIPLICATION is the next basic operation that you will need to know for the AFOQT. The multiplication of two numbers, which is basically repeated addition, will result in a number that is known as the product. The very first step in multiplication is putting the number with the most digits on top and lining up the numbers to the right. There are two general situations you will encounter when you have to multiply two numbers together.

1. One of the numbers you have to multiple has only a single digit. For example multiplying 15 by 8.
2. Both of the numbers you have to multiply have more than one digit. For example multiplying 10 by 16.

In the first situation, you want to put the number with the single digit on the bottom and then multiply all of the digits of the top number by the bottom number. You will begin on the right-hand side of the top number and then multiply individual digits, working to the left. If the multiplication results in a number with more than one digit, you will write down the ones digit in the product and carry the tens digit to the next column. Then you will add the carried number to the next multiplication.

EXAMPLE

Multiply 123 by 6.

123

× 6

Notice how the number with the most digit is put on top. The first step would be to look at the singles digit (6). Multiple that number (6) by the rightmost number in the top column (3). That will result in 18, so we'll put an 8 below. Moving left, the 6 is multiplied by 2 (on a diagonal), resulting in 12, but we also have to add the 1 that we just carried, so we get 13. Put a 3 below, and carry 1. Now the 6 is multiplied by the 1 to get 6, but we need to carry over the 1 from the 13, so we get 7. The final result is 738.

123

× 6

738

The second situation is the same as the first, but with an additional step. Again, remember to place the number with the most digits on top, and start multiplying from the right (the ones digits). You will first multiple the ones column by the top number, then the tens, then the hundreds, and so on until you are finished. Each value place you move to the left will add a zero to the right of the product. The products will then be added together to give the final result.

Multiply 201 by 13.

$$
\begin{array}{r}
201 \\
\times\ 13 \\
\hline
603 \\
+\ 2010 \\
\hline
\mathbf{2613}
\end{array}
$$

The first thing that has so be done after arranging the numbers in the correct way is to multiply the 3 by 1. That will give you 3. Then 3 by 0 and 3 by 2 respectively, for a result of 603 total. Now you will do the same thing with the 1 (the tens digit of the second number). 1 by 201 is, obviously, 201. But since you are using the tens digit, add a 0 to the right-hand side of this number, giving you 2010. This is to reflect the higher value of the digit being multiplied. Now, finally, you will add the 603 and 2010, giving you a final result of **2613.**

The multiplication of decimals requires something quite different than what you did during the addition and subtraction. In this situation, you will completely ignore the decimal point altogether. Multiply the two numbers together as if they were whole numbers. Once you are done doing that, you will figure out where to place the decimal point back in. This sounds much more difficult than it actually is.

Multiply 54.8 by 9.9.

$$
\begin{array}{r}
54.8 \\
\times\ 9.9 \\
\hline
\end{array}
$$

First, you will take out the decimal points so that the multiplication will be a bit simpler.

$$
\begin{array}{r}
548 \\
\times\ 99 \\
\hline
\end{array}
$$

Now finish up the multiplication and addition:

$$
\begin{array}{r}
548 \\
\times\ 99 \\
\hline
4932 \\
+\ 49320 \\
\hline
54252
\end{array}
$$

To deal with the decimal points, add up the number of digits to the right of the decimal points in the original numbers. This does not mean to add up the actual numbers; it means add the number of total digits to the right of the decimal points. So 54.8 has one digit to the right of the decimal point and 9.9 has one digit to the right of the decimal point. This means you have 1 + 1 digits to the right of the decimal points (2). In the final product, 54252, just stick the decimal point in (from the right) based on that number. The result would be **542.52**.

This process is a bit more complicated than the simple addition and subtraction that was done earlier and it is also more complicated than the multiplication of whole numbers. Regardless, it will get a lot easier once you have done it a few times.

Division

DIVISION is the process of splitting a number into parts or groups, and this yields a quotient. There are four basic steps that you will undertake to do long division and you will repeat these steps for every division operation that you have to perform.

EXAMPLE

Divide 5 into 65.

The first step here is going to be to set up the problem:

$\frac{65}{5}$ or $5\sqrt{65}$ fd

Now you will begin the actual division. First, select the leftmost digit of the number being divided (the dividend) and see how many times the divisor goes into it. In this case, 5 goes into 6 one (1) time. 6 – 5 is 1, leaving a remainder of 1. Now you will take the remainder, along with the next digit in the number being divided, in this case 1 and 5 respectively, and see how many times the divisor goes into it. How many times does 5 go into 15? The answer is **3**. Pair that with the original digit you got and you have your solution: 13.

```
    1 3
5√ 65
    5 0
    1 5
    1 5
      0
```

To test this and see if you got the right answeryour answer, multiply 13 by 5 and see what you get. Is it the original number, 65? Yes, it is. So the division is correct. Always test once this has been done, and remember to add any remainder you might get to the product of the divisor and dividend when checking your answer.

When doing this with decimals, you will work exactly as you did with multiplication. Ignore the decimals until the end and then deal with them just like before. Add the decimal in exactly where it would be in the number that is being divided (move the decimal straight up into the answer). However, if there is a decimal in the outside (divisor), you must move the decimal to the right to make this a whole number, and then do the same (same number of places) on the inside (the dividend). Then you move the decimal straight up.

Word Problems

Basic Problems

The most basic problems that you will encounter on the arithmetic reasoning section of the AFOQT are problems that require one or two steps to complete. Generally, the most difficult part about these types of problems is figuring out what they are actually asking you to do.

One or two step problems usually have only one computation that needs to be made. At times, you may have to do problems that are a bit more complex. With that being said, the question you are being asked will usually be clear and not hard to figure out. The way these problems will be discussed and taught here will involve logic and examples. The exact computations will be up to you to complete on your own. And take advantage of the fact that the answers are provided by plugging them in backwards, if you need to.

EXAMPLE

1. Sally is going to the store and she is buying her groceries for the week. She is buying some apples for $5.00, some pears for $4.25, and some grapes for $9.50. How much does she spend at the store during this trip?

    This is a very simple problem. It is one basic step and only involves basic arithmetic. Three numbers are provided: $5.00, $4.25, and $9.50. The question asks you how much she spends. The way to solve this is to sum the three numbers, giving you a total of **$18.75**.

2. John has a job laying tile with his father's company. He is paid $9 an hour. He works for two weeks per pay period. This period, he works 10 hours during the first week and 5 hours during the second week. How much did John make during this pay period?

    This problem is a bit more complex since two operations are used. The information you are given lays it all out: He gets $9 an hour, a pay period is 2 weeks, he worked 10 hours the first week and 5 hours the second week. That's 15 total hours. 15 hours times $9 per hour yields $135 for this two-week pay period. You can also solve this by multiplying 10 hours by $9 and then 5 hours by $9 and adding them together. The result will be the same, even though the process is different.

Percent and Interest

The most common types of problems that you will encounter in the real world (in terms of word problems for mathematics) are going to be dealing with money. This is when **PERCENT AND INTEREST** operations come into play. Either you will be dealing with interest that is being accrued on some amount of money in the bank or on a loan you are paying, or you will be trying to find a certain percentage of a sum. These two things are typically related.

Here are two things to keep in mind when you begin working on this type of problem:

- The first thing you need to do is to convert the percentage into a whole number or decimal. You can do this by removing the percentage and moving the decimal two places to the left. (Remember, when you are removing the percent, move the decimal away from it.)
- The second thing you need to do is multiply the decimal (originally the percentage) by the figure being calculated.
- Finally, if necessary, add the percentage amount (which is now a whole number or decimal) to the original amount.

EXAMPLE

You have an investment account and you have put in $6,000. The account earns 7.5% interest annually. How much money do you have, in total, when one year is up?

The first thing you want to do is determine the important numbers here. Beginning amount = 6000, 1 year = 7.5% interest. Interest interest for one year = 7.5% (or 0.075).

You can do this in one of two ways. You can set it up like a standard multiplication problem and then do the addition, or you can set it up algebraicallyin one step. Here is each method in fullthe first way:

$6000 × 0.075 = 450

$6000 + $450 = $6450

In this method, you do the multiplication to get the interest added for one year, and then add the total of the interest for one yearthis to the original sum.

The second method involves a bit of algebrasaves a step by combining the interest and the original amount:. In this one, t = time (in years).

$6000 × 1.075 = **$6450**

Again, when you are dealing with percentages, it is important to convert them in the proper way by moving the decimal two places to the left when removing the percentage; for example, 10% is 0.10 and

5% is 0.05. Percentages between 1% and 99% take up the first two digits to the right of the decimal point in a conversion.

Ratios and Proportions

RATIOS involve comparisons between numbers, and PROPORTIONS involve the equalities of ratios. Ratios are usually written with a ":" separating them. So a 5 to 1 ratio would be written 5:1. These can be written fractionally as well, which is useful when you begin working with some of the word problems that involve actual proportions. For example, 5:1 would be $\frac{5}{1}$. Comparing two numbers using a ratio results (with a 1 on the bottom typically) is what is known as a unit rate. This is where you get terms like "miles per hour" or "kilometers per hour" since "per" can mean "divided by"(this is the same as finishing the division of a fractional form of a ratio).

EXAMPLE

You want to figure out how much you are paying for your flour. On the price tag, it says that the flour is $9.95 and that the bag contains 6 ounces of flour. How much does the flour cost on a per ounce basis?

You want the cost of one ounce of flour, but you are given the cost of six ounces. Turn this rate into a fraction, and then set up a proportion with ounces on the left and money on the right: $\frac{1}{6} = \frac{x}{9.95}$. You can cross-multiply to get $6x = 9.95$, so the final result is **$1.66** (rounded to the nearest cent).

All problems of this type will be solved in roughly the same way, as you will see when you begin doing them for yourself.

Problems of Measurement

Measurement problems are extremely simple. You are going to, almost universally, be asked to determine either the area or the perimeter of a given shape. The perimeter is the distance around the geometric shape. The area is the total amount of space inside of the geometric shape.

This is the formula for finding perimeter, where l = length, w = width, and P = perimeter:

$$P = l + l + w + w$$

That can also be written as $P = 2l + 2w$. It is both of the lengths added with both of the widths. The formula for area, where l = length, w = width, A = area is:

$$A = l \times w$$

Area is length times width.

Tips

Here are some tips to help you with the arithmetic reasoning subtest of the AFOQT:

- Make sure you read the entire question before you attempt to answer. It is extremely important to confirm what the question is actually asking you, since with multiple choice problems, they will try to trick you.

- The answer should, at a glance, make sense based on what you are asked for. If you are asked how many eggs someone is carrying home from the store, the answer is going to likely be in the range of 0 – 100. So you can quickly eliminate any ridiculous looking answers early on.

- When you set up the equation that you will be using, be careful to put the correct numbers in the correct places. For example, when setting up proportions, make sure matching units are in the same spots (for example, on the left or on top).

- Generally, they will not give you figures if you don't need to use them. So if you see measurements and numbers in the problem, be sure to take note and figure out where you need to use them.

- Follow the order of operations. This shouldn't really even have to be said because it is so obvious, but always follow it. Some of the answers will likely be designed to trick you as if you hadn't done the math in the correct order.

- When it comes to basic arithmetic operations, you will find that practice makes perfect. The more you can practice, the easier the math problems will be. Make sure units match; for example, you may gave to convert some of the numbers from hours to minutes to match other numbers.

- If you can't figure out answer from the question, try each answer in the multiple choice to see if they work. In other words, you may have to work backwards for some problems.

- Try simpler numbers if you can't figure out the problems with the numbers given. For example, use $100 for percent problems to figure out the correct operation(s). Then apply these operations, using the more complex numbers in the problem.

- For word problems with variables representing real numbers, use real numbers to get an answer (to see what you're doing) and then put variables back in.

- Draw pictures!

Practice Questions

1. Which of the following two integers have a sum equaling a number less than 149?

   1-A   89, 65

   1-B   55, 93

   1-C   93, 57

   1-D   84, 77

2. John is going to a big party tonight and he needs to buy a new dress shirt for the occasion. The last time he went to the store, he bought one for around $20. The price is now $28. What percentage of the original price has it increased?

   2-A   30%

   2-B   40%

   2-C   45%

   2-D   50%

3. Rebecca, Emily, and Kate all live on the same straight road. Rebecca lives 1.4 miles from Kate and 0.8 miles from Emily. What is the minimum distance Emily could live from Kate?

   3-A   2.2 miles

   3-B   1.1 miles

   3-C   0.8 miles

   3-D   0.6 miles

4. One of the classes in a local elementary school has g girls. That is three more than four times how many boys are in that class. How many boys are in the local elementary school class? You may use x to represent the number of boys.

   4-A $x = \frac{g-3}{4}$

   4-B   13

   4-C   12

   4-D $b = \frac{4}{g-3}$

5. One number, y, is 10 more than a second number. If you double that second number and add triple the higher number, the resulting number is 55. What are those two numbers?

   5-A   2, 12

   5-B   10, 12

   5-C   5, 15

   5-D   5, 10

6. Convert 28% to a decimal.

   6-A   2.8

   6-B   0.028

   6-C   0.0028

   6-D   0.28

7. A man has a set of jacks. He has j jacks, which is twice as many jacks as he has balls (b). How many balls does he have?

   7-A $b = 2j$

   7-B $b = \frac{j-4}{2}$

   7-C $j = \frac{b-4}{2}$

   7-D $b = \frac{j}{2}$

8. A class has around 30 students in it. Those students are divided into three groups. The first group has the most students in it. The second and third group both have 8 students each. How many students are in the first group?

   8-A   14

   8-B   15

   8-C   8

   8-D   16

9. John is a tractor driver. He works for $20 an hour and he gets paid an additional $5 for every acre of land he works. John is hired to work on a small farm. The work takes him 6 hours and he manages to cover all 10 acres of the farm. How much money is John going to be paid for this job?

9-A $120

9-B $50

9-C $150

9-D $170

10. You are doing some yardwork and you need to put seed onto your lawn. One bag of seed will cover 15 square feet. How many bags are you going to need in order to cover a yard that is 5 feet by 6 feet?

10-A 2 bags

10-B 3 bags

10-C 5 bags

10-D 15 bags

11. You are buying supplies for your class cookout. You are expecting 32 students and 2 teachers to attend. You estimate that for every 3 people, you will need 5 hamburgers. The hamburger patties come in packs of 6. How many packs should you buy?

11-A 10 packs

11-B 9 packs

11-C 8 packs

11-D 4 packs

12. Mark has a yard that is 18 feet long by 30 feet wide. What is the perimeter of the yard that Mark has?

12-A 48 feet

12-B 96 feet

12-C 540 feet

12-D 12 feet

13. A large construction company is trying to lay concrete for a new apartment building. They charge by the square foot. The price per square foot is 35 cents. The apartment building is going to be 500 feet long by 325 feet wide. How much is the total cost going to be for laying the concrete?

13-A $56,875

13-B $170,625

13-C $16,350

13-D $162,500

14. Sally is working a new job. She gets paid $10 per hour for stocking the shelves at a local grocery store. She is paid weekly and this is the week that her bills are due for the month. Her rent is $100, her car payment is $50, and her groceries cost her $35. She worked 40 hours this week. How much money does she have left over in her paycheck after she pays all of her bills?

14-A $394

14-B $185

14-C $400

14-D $215

15. You are looking to buy a home. The price of the home is $135,000. Every year, you will have to pay 2.5% of the value of the home in taxes. After 3 years, how much money will you have paid in taxes on the home?

15-A $3,375

15-B $13,375

15-C $10,125

15-D $6750

Arithmetic Reasoning Answer Key

1.	B.	9.	D.
2.	B.	10.	A.
3.	D.	11.	A.
4.	A.	12.	B.
5.	C.	13.	A.
6.	D.	14.	D.
7.	D.	15.	C.
8.	A.		

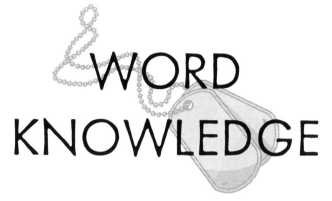

three

WORD KNOWLEDGE

The military considers clear and concise communication so important that it is taught and graded at all levels of leadership training. If you are planning a military career, you will be tested on your verbal skills as you move through the ranks.

The good news is that most individuals have been exposed to all of the vocabulary words used on the subtest by the time they have reached the tenth grade. This does not mean that you are going to recognize every single word. It does mean, however, that you won't be expected to know advanced Latin or graduate science terminology.

This section of the test gives you twenty-five questions to answer in five minutes. This may seem like a disproportionate amount of time—it comes out to about twelve seconds per question—but don't worry! We're going to arm you with all of the knowledge you'll need to work quickly and efficiently through this section.

As an extra challenge to this section, the questions are formatted, so the word has to be matched without context. You will typically be given a single word in all capital letters, and then you must choose from the answer choices that word matches or has the same meaning. An example is shown below.

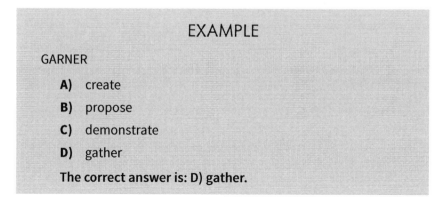

EXAMPLE

GARNER

  A) create

  B) propose

  C) demonstrate

  D) gather

The correct answer is: D) gather.

You will note that as we work through this section, not all questions will be formatted this way and that is for a reason. Words can be tricky and there is no shortcut to learning to understand them, or even refreshing what you might have learned years ago. The biggest mistake one can make is thinking they will just memorize a bunch of vocabulary words. This is a wasted effort. First of all, it is very unlikely you will retain that information long enough from rote memorization. Secondly, it would take endless hours to memorize enough of them. Finally, you would still be missing the key elements that allow you to figure out the meaning of a word. Without knowing its definition OR knowing what it certainly does not mean allows you to eliminate wrong answer choices.

Vocabulary Basic Training

The first step in getting ready for this section of the AFOQT consists of reviewing the basic techniques used to determine the meanings of words you are not familiar with. The good news is that you have been using various degrees of these techniques since you began to speak. Sharpening these skills will help you with the paragraph comprehension subtest.

Following each section you will find a practice drill. Use your scores on these to determine if you need to study a particular subject matter further. At the end of each section, you will find a practice drill to test your knowledge.

The questions found on the practice drills are not given in the two formats found on the Word Knowledge subtest; rather they are designed to reinforce the skills needed to score well on the Word Knowledge subtest.

Context Clues

Although you won't get any context to reference words on the AFOQT, for training purposes we will start with words in context, so you can start to see how to break down a word into its meaning. Your ability to observe sentences closely is extremely useful when it comes to understanding new vocabulary words.

Types of Context

There are two different types of context that can help you understand the meaning of unfamiliar words: sentence context and situational context. Regardless of which context is present, these types of questions are not testing your knowledge of the vocabulary; they are testing your ability to comprehend the meaning of a word through its usage.

SITUATIONAL CONTEXT is the basis of the Paragraph Comprehension subtest and will be discussed in chapter two.

Sᴇɴᴛᴇɴᴄᴇ ᴄᴏɴᴛᴇxᴛ occurs in the sentence containing the vocabulary word. To figure out words using sentence context clues, you should first determine the most important words in the sentence.

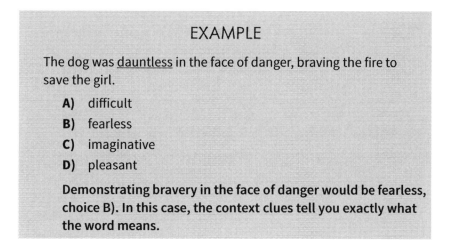

EXAMPLE

I had a hard time reading her <u>illegible</u> handwriting.

- **A)** neat
- **B)** unsafe
- **C)** sloppy
- **D)** educated

Already, you know that this sentence is discussing something that is hard to read. Look at the word that *illegible* is describing: *handwriting*. Based on context clues, you can tell that illegible means that her handwriting is hard to read.

Next, look at the choices. Choice A) neat is obviously wrong because neat handwriting would not be difficult to read. Choice B) unsafe and D) educated don't make sense. Therefore, choice C) sloppy is the best answer choice.

Types of Clues

There are four types of clues that can help you understand the context, which in turn helps you define the word. They are restatement, positive/ negative, contrast, and specific detail.

Rᴇsᴛᴀᴛᴇᴍᴇɴᴛ ᴄʟᴜᴇs occur when the definition of the word is clearly stated in the sentence.

EXAMPLE

The dog was <u>dauntless</u> in the face of danger, braving the fire to save the girl.

- **A)** difficult
- **B)** fearless
- **C)** imaginative
- **D)** pleasant

Demonstrating bravery in the face of danger would be fearless, choice B). In this case, the context clues tell you exactly what the word means.

Pᴏsɪᴛɪᴠᴇ/ɴᴇɢᴀᴛɪᴠᴇ ᴄʟᴜᴇs can tell you whether a word has a positive or negative meaning.

GO ON

EXAMPLE

The magazine gave a great review of the fashion show, stating the clothing was <u>sublime</u>.

A) horrible

b) exotic

c) bland

d) gorgeous

The sentence tells us that the author liked the clothing enough to write a great review, so you know that the best answer choice is going to be a positive word. Therefore, you can immediately rule out choices A) and C) because they are negative words. Exotic is a neutral word; alone, it doesn't inspire a great review. The most positive word is gorgeous, which makes choice D) the best answer.

The following sentence uses both restatement and positive/negative clues:

> "Janet suddenly found herself <u>destitute</u>, so poor she could barely afford to eat."

The second part of the sentence clearly indicates that destitute is a negative word; it also restates the meaning: very poor.

CONTRAST CLUES include the opposite meaning of a word. Words like *but*, *on the other hand*, and *however* are tip-offs that a sentence contains a contrast clue.

EXAMPLE

Beth did not spend any time preparing for the test, but Tyron kept a <u>rigorous</u> study schedule.

A) strict

B) loose

C) boring

D) strange

In this case, the word but tells us that Tyron studied in a different way than Beth. If Beth did not study very hard, then Tyron did study hard for the test. The best answer here, therefore, is choice A) strict.

SPECIFIC DETAIL CLUES give a precise detail that can help you understand the meaning of the word.

It is important to remember that more than one of these clues can be present in the same sentence. The more there are, the easier it will be to determine the meaning of the word, so look for them.

Denotation and Connotation

As you know, many English words have more than one meaning. For example, the word *quack* has two distinct definitions: the sound a duck makes, and a person who publicly pretends to have a skill, knowledge, education, or qualification that they do not possess.

The **DENOTATION** of a word is the dictionary definition.

The **CONNOTATION** of a word is the implied meaning(s) or emotion that the word makes you think. For example:

> "Sure," Pam said excitedly, "I'd just love to join your club; it sounds so exciting!"

> "Sure," Pam said sarcastically, "I'd just love to join your club; it sounds so exciting!"

Even though the two sentences only differ by one word, they have completely different meanings. The difference, of course, lies in the words *excitedly* and *sarcastically*.

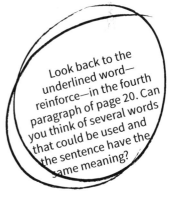

Look back to the underlined word—reinforce—in the fourth paragraph of page 20. Can you think of several words that could be used and the sentence have the same meaning?

Roots, Prefixes, and Suffixes

You just got done with what could be called a warm-up exercise, and now we will get into the tougher material you need to know specifically for the AFOQT. Although you are not expected to know every word in the English language, you are expected to have the ability to use deductive reasoning to find the choice that is the best match for the word in question. This is why we are going to explain how to break a word into its parts of meaning:

PREFIX – ROOT – SUFFIX

One trick in dividing a word into its parts is first to divide the word into its SYLLABLES. To show how syllables can help you find roots and affixes, we'll use the word DESCENDANT, which means one who comes from an ancestor. Start by dividing the word into its individual syllables; this word has three: *de-scend-ant*. The next step is to look at the beginning and end of the word, and then determine if these syllables are prefixes, suffixes, or possible roots. You can then use the meanings of each part to guide you in defining the word. When you divide words into their specific parts, they do not always add up to an exact definition, but you will see a relationship between their parts.

It's important to note that this trick won't work in every situation, because not all prefixes, roots, and suffixes have only one syllable. For example, take the word monosyllabic (which ironically means one syllable). There are five syllables in that word, but only three parts. The prefix is *mono*, meaning one. The root *syllab* refers to *syllable*, while the suffix *ic* means pertaining to. Therefore, we have one very long word which means pertaining to one syllable.

The more familiar you become with these fundamental word parts, the easier it will be to define unfamiliar words. Although the words found on the Word Knowledge subtest are considered vocabulary words learned by the tenth grade level of high school, some are still less likely to be found in an individual's everyday vocabulary. The root and affixes list in this chapter uses more common words as examples to help you learn them more easily. Don't forget that you use word roots and affixes every day, without even realizing it. Don't feel intimidated by the long list of roots and affixes (prefixes and suffixes) at the end of this chapter. You already know and use them every time you communicate with some else, verbally and in writing. If you take the time to read through the list just once a day for two weeks, you will be able to retain most of them and understand a high number of initially unfamiliar words.

Roots

Roots are the building blocks of all words. Every word is either a root itself or has a root. Just as a plant cannot grow without roots, neither can vocabulary because a word must have a root to give it meaning. For example:

The test instructions were <u>unclear</u>.

The root is what is left when you strip away all the prefixes and suffixes from a word. In this case, take away the prefix *un-*, and you have the root *clear*.

Roots are not always recognizable words because they come from Latin or Greek words, such as *nat*, a Latin root meaning born. The word *native*, which means a person born of a referenced placed, comes from this root, as does the word *prenatal*, meaning before birth. Yet, if you used the prefix *nat* instead of *born*, just on its own, no one would know what you were talking about.

Words can also have more than one root. For example, the word *omnipotent* means all powerful. Omnipotent is a combination of the roots *omni-*, meaning all or every, and *-potent*, meaning power or strength. In this case, *omni* cannot be used on its own as a single word, but *potent* can. Again, it is important to keep in mind that roots do not always match the exact definitions of words. They can have several different spellings, but breaking a word into its parts is still one of the best ways to determine its meaning.

Prefixes and Suffixes

Prefixes are syllables added to the beginning of a word and suffixes are syllables added to the end of the word. Both carry assigned meanings. The common name for prefixes and suffixes is AFFIXES. Affixes do not have to be attached directly to a root and a word can often have more than one prefix and/or suffix. Prefixes and suffixes can be attached to a word to change completely the word's meaning or to enhance the word's original meaning. Although they don't mean much to us on their own, when attached to other words affixes can make a world of difference.

We can use the word prefix as an example:
- FIX means to place something securely.
- PRE means before.
- PREFIX means to place something before or in front.

An example of a suffix:
- FEMIN is a root. It means female, woman.
- -ISM means act, practice or process.
- FEMINISM is the defining and establishing of equal political, economic, and social rights for women.

Unlike prefixes, suffixes can be used to change a word's part of speech.

For example, take a look at these sentences:

Randy raced to the finish line.

Shana's costume was very racy.

In the first sentence, *raced* is a verb. In the second sentence, *racy* is an adjective. By changing the suffix from *-ed* to *-y*, the word *race* changes from a verb into an adjective, which has an entirely different meaning.

Although you cannot determine the meaning of a word by a prefix or suffix alone, you can use your knowledge of what root words mean to eliminate answer choices. Indicate if the word is positive or negative and you will get a partial meaning of the word.

Synonyms and Antonyms

SYNONYMS are groups of words that mean the same, or almost the same, thing as each other. The word *synonym* comes from the Greek roots *syn-*, meaning same, and *-nym*, meaning name. Hard, difficult, challenging, and arduous are synonyms of one another.

ANTONYMS are sets of words that have opposite, or nearly opposite, meanings of one another. The word *antonym* comes from the Greek roots *ant-*, meaning opposing, and *-nym* meaning name. Hard and easy are antonyms.

Synonyms do not always have exactly the same meanings, and antonyms are not always exact opposites. For example, *scalding* is an adjective that means burning. Boiling water can be described as scalding or as hot. *Hot* and *scalding* are considered synonyms, even though the two words do not mean exactly the same thing; something that is scalding is considered to be extremely hot.

In the same manner, antonyms are not always exact opposites. *Cold* and *freezing* are both antonyms of scalding. Although freezing is closer to being an exact opposite of scalding, cold is still considered an antonym. Antonyms can often be recognized by their prefixes and suffixes.

Here are rules that apply to prefixes and suffixes of antonyms:

- Many antonyms can be created simply by adding prefixes. Certain prefixes, such as *a-*, *de-*, *non-*, and *un-*, can be added to words to turn them into antonyms. *Atypical* is an antonym of typical, and *nonjudgmental* is an antonym of judgmental.
- Some prefixes and suffixes are antonyms of one another. The prefixes *ex-* (out of) and *in-/il-/im-/ir-* (into) are antonyms, and are demonstrated in the antonym pair exhale/inhale. Other prefix pairs that indicate antonyms include *pre-/post-*, *sub-/super-*, and *over-/under-*. The suffixes *-less*, meaning without, and *-ful*, meaning full of, often indicate that words are antonyms as well. For example: *meaningless* and *meaningful* are antonyms.

Review

Remember that roots are the basic unit of meaning in words. When you read a word that is unfamiliar to you, divide the word into syllables and look for the root by removing any prefixes and suffixes.

You have also learned that prefixes and suffixes are known collectively as **AFFIXES**. Although affixes are not words by themselves, they are added to roots or words to change the meaning of roots or change a word's part of speech. **PREFIXES** change or enhance the meanings of words, and are found at the beginning of words. **SUFFIXES** change or enhance the meanings of words and/or change parts of speech and are found at the end of words.

You have also learned that **SYNONYMS** are words that have the same or almost the same meaning, while **ANTONYMS** are words that have opposite or nearly opposite meanings. Synonyms and antonyms of a word will always share the same part of speech. That is, a synonym or antonym of a verb has to be a verb; a synonym or antonym of an adjective has to be an adjective; and so forth. We also learned that not all words have synonyms or antonyms, and that synonyms do not always have exactly the same meaning, just as antonyms do not have to be exact opposites.

Tips

Use words that you are very familiar with as examples when you study word roots. The more familiar the word is to you, the easier it will be for you to remember the meaning of the root word. Use words that create a vivid picture in your imagination.

Be sure to look at all parts of the word to determine meaning.

Remember the power of elimination on an exam. Use your knowledge of word roots to eliminate incorrect answers. The more you narrow down your choices, the better your chances of choosing the correct answer. You have to do so quickly of course, but even eliminating one wrong answer before guessing greatly increases your chances.

Roots do not always match the exact definitions of words. Another important thing to keep in mind is that sometimes one root will have several different spellings.

Affixes do not have to be attached directly to a root. A word can often have more than one affix, even more than one prefix or suffix. For instance, the word *unremarkably* has two prefixes (*un-* and *re-*) and two suffixes (*-able* and *-ly*).

GO ON

Table 3.1. Roots and definitions

ROOT	DEFINITION	EXAMPLE
ast(er)	star	asteroid, astronomy
audi	hear	audience, audible
auto	self	automatic, autograph
bene	good	beneficent, benign
bio	life	biology, biorhythm
chrono	time	chronometer, chronic
dict	say	dictionary, dictation
duc	lead or make	ductile, produce
gen	give birth	generation, genetics
geo	earth	geography, geometry
graph	write	graphical, autograph
jur or jus	law	justice, jurisdiction
log or logue	thought	logic, logarithm
luc	light	lucidity
man	hand	manual
mand	order	remand
mis	send	transmission
path	feel	pathology
phil	love	philanthropy
phon	sound	phonograph
port	carry	export
qui	quiet	quiet
scrib or script	write	scribe, transcript
sense or sent	feel	sentiment
tele	far away	telephone
terr	earth	terrace
vac	empty	vacant
vid	see	video
vis	see	vision
omni	all	omnivores
cap	take	capture
ced	yield	secede
corp	body	corporeal
demo	people	democracy
grad	step	graduate
crac or crat	rule	autocrat
mono	one	monotone
uni	single	Unicode
ject	throw	eject

Practice Problems

1. They investigated the <u>alleged</u> human rights violations.

   1-A proven

   1-B false

   1-C unproven

   1-D horrific

2. CEDE

   2-A consign

   2-B surrender

   2-C keep

   2-D abandon

3. AFFLICT

   3-A attack

   3-B perturb

   3-C assist

   3-D agonize

4. CONSPICUOUS

   4-A bold

   4-B unremarkable

   4-C quiet

   4-D dull

5. <u>Insurgents</u> were responsible for a number of attacks, including suicide bombings.

   5-A anarchists

   5-B communists

   5-C rebels

   5-D patriots

6. AUSTERE

   6-A welcoming

   6-B ornate

   6-C simple

   6-D fanciful

7. ADMONISH

   7-A denounce

   7-B dislike

   7-C reprimand

   7-D praise

8. DEFERENCE

   8-A defiance

   8-B submissiveness

   8-C hostility

   8-D sociability

9. The site had been <u>neglected</u> for years.

   9-A ignored

   9-B maintained

   9-C crumbling

   9-D growing

10. INSINUATE

    10-A imply

    10-B introduce

    10-C proclaim

    10-D abbreviate

11. EXPLICATE

    11-A obscure

    11-B decipher

    11-C clarify

    11-D confuse

12. DECORUM

    12-A propriety

    12-B decoration

    12-C drunkenness

    12-D indecency

GO ON

13. He was <u>chagrined</u> when he tripped and fell in the hallway.

   13-A   injured

   13-B   embarrassed

   13-C   unharmed

   13-D   angry

14. AUDACIOUS

   14-A   frightening

   14-B   engaging

   14-C   daring

   14-B   boring

15. The <u>intrepid</u> volunteers worked in the refugee camps.

   15-A   uncaring

   15-B   caring

   15-C   compassionate

   15-D   fearless

16. SURREPTITIOUS

   16-A   hidden

   16-B   clandestine

   16-C   public

   16-D   illegal

17. To take precaution is to:

   17-A   prepare before doing something

   17-B   remember something that happened earlier

   17-C   become aware of something for the first time

   17-D   try to do something again

18. To reorder a list is to:

   18-A   use the same order again

   18-B   put the list in a new order

   18-C   get rid of the list

   18-D   find the list

19. An antidote to a disease is:

   19-A   something that is part of the disease

   19-B   something that works against the disease

   19-C   something that makes the disease worse

   19-D   something that has nothing to do with the disease

20. Someone who is multiethnic:

   20-A   likes only certain kinds of people

   20-B   lives in the land of his or her birth

   20-C   is from a different country

   20-D   has many different ethnicities

21. Someone who is misinformed has been:

   21-A   taught something new

   21-B   told the truth

   21-C   forgotten

   21-D   given incorrect information

22. Awe is most dissimilar to:

   22-A   contempt

   22-B   reverence

   22-C   valor

   22-D   distortion

23. Intricate is most similar to:

   23-A   delicate

   23-B   costly

   23-C   prim

   23-D   complex

24. Skeptic is most dissimilar to:

   24-A   innovator

   24-B   friend

   24-C   politician

   24-D   believer

25. Hypothetical is most dissimilar to:

    25-A   uncritical

    25-B   actual

    25-C   specific

    25-D   thoughtful

GO ON

Word Knowledge Answer Key

1.	C	14.	C
2.	B	15.	D
3.	D	16.	B
4.	A	17.	A
5.	C	18.	B
6.	C	19.	B
7.	C	20.	D
8.	B	21.	D
9.	A	22.	A
10.	A	23.	D
11.	C	24.	D
12.	A	25.	B
13.	B		

four

MATHEMATICS KNOWLEDGE

Introduction

The purpose of the **MATHEMATICS KNOWLEDGE** section on the test is to make sure you fully understand the concepts that are important in high school mathematics courses. This includes information about basic operations, order of operations, algebra, and geometry. It also includes working with fractions, which can prove difficult for many people.

Number Theory

NUMBER THEORY is the study of the properties of whole numbers (0, 1, 2, …) and also integers, which are whole numbers plus their negative counterparts. Negative numbers can be thought of as inverses or opposites of whole numbers. Integers, like whole numbers, can be written without the use of a fractional part.

Prime Numbers

Prime numbers are numbers that have only two factors, 1 and itself. Examples would include 2, 3, 5, 7, 11, 13, 17, 19, and 23, as well as many others. When you are attempting to figure out if a number is a prime number or not, all you need to do is figure out whether other numbers, besides 1 and the number itself, will divide evenly into it.

EXAMPLE

Is 66 a prime number?

66 is not going to be considered a prime number. It can be divided as 2 and 33, as 11 and 6, or as 1 and 66.

It is important to note that prime numbers are usually going to be odd, with the exception of the number 2. Even numbers clearly can be divided by 2, so it won't work for larger numbers.

Mean (Average), Standard Deviation, Median, and Mode

You may be asked to find the MEAN, STANDARD DEVIATION, MEDIAN, or MODE of a set of numbers. The mean (average) of numbers is obtained by adding up all the numbers and then dividing by the number of numbers that you added up. The standard deviation is a measurement (which you probably won't have to compute!) of how far apart the numbers are from the mean or average. The median is obtained by ordering the numbers and picking the middle number, or averaging the two middle numbers. The mode is obtained by finding the number repeated the most number of times (if there is a number that repeats).

Multiples

MULTIPLES of numbers are what results from multiplying whole numbers by other numbers. For example, multiples of 7 are 7, 14, 21, and so on.

Common multiples of two numbers are the numbers that are multiples of both. If you were looking at, for example, 4 and 8, then 16 would be a common multiple. 8 would also be a common multiple of the two numbers.

The least common multiple is the smallest common multiple that two given numbers share. The fastest way to find this is to just write out the first few multiples for both numbers that you have available and then figure out which one is the smallest. This is not a particularly difficult task, but you will have to understand how multiples work in order to do it properly. For example, the least common multiple of 4 and 8 is 8.

Factors

A FACTOR is a number that goes evenly into another number with no remainder. So think of factors as all the numbers you can multiply together to get another number. Here is an example:

EXAMPLE

What are the factors of 24?

1 and 24, 2 and 12, 3 and 8, 4 and 6, since all of these numbers go into 24 exactly (without a remainder).

It is not necessary to write them like that, but it will help you keep them in mind if you keep the factors together with their counterparts. Another way to write these is 1, 2, 3, 4, 6, 8, 12, and 24.

If you want to figure out whether a number is a factor of another one or not, just divide the number by the potential factor and see if the result is a whole number. Every number will have at least two factors, 1 and the number itself. For example, the two factors of 2 are 1 and 2.

Numbers that are the factors of more than one whole number are known as **COMMON FACTORS**. For example, 6 would be a common factor of both 12 and 24 (because it can be multiplied by 2 to reach 12 and 4 to reach 24). The largest factor that goes into two numbers is called the greatest common factor.

Exponents

EXPONENTS and exponential notation are used to help simply expressions, particularly when factors are repeated multiple times. Exponents are written as superscripts above the number that has the exponent. The number that has the exponent is called the base, so 8 is the base and 2 is the exponent in this example:

$$8 \times 8 = 8^2$$

Think of exponents as shorthand that is used to help keep the math straight and stop it from becoming too confusing. This can be seen below:

$$5 \times 5 \times 5 \times 5 \times 5 \times 5 = 5^6$$

Writing the number as an exponent makes it much easier to see what is happening and will simplify equations for you. Another way to say what is happening above is "five to the sixth power" or "five to the sixth". Here is another example where we are simplifying with an exponent:

$$2 + 2 \times 5 \times 5 \times 6 \times 7 + 8 = 2 + 2 \times 5^2 \times 6 \times 7 + 8$$

Does it simplify that equation a lot? Not in this case, but it does make it a little bit simpler to read.

Exponents are also used to denote cumbersome numbers in scientific notation; for example, $54{,}000 = 5.4 \times 10^4$ and $.0043 = 4.3 \times 10^{-3}$.

Let's say a few words about fractional exponents and negative exponents. In an exponent that is a fraction, the number on the top acts just like an exponent, but the number on the bottom designates a "root" (see next section), which means a number would have to multiplied that many times to get to that number. For example, $8^{\frac{2}{3}} = (\sqrt[3]{8^2}) = (\sqrt[3]{8})^2 = 2 \times 2 = 4$. With negative exponents, you have to take the reciprocal of the base and make the exponent positive. For example, $2^{-3} = \left(\frac{1}{2}\right)^3 = \frac{1}{8}$.

Also note that you can use exponents for questions dealing with combinations of letters or numbers. For example, if you were asked how

The phrase *simplify the expression* just means you need to perform all the operations in the expression.

many different numbers on a license plate there could be if the first three digits were letters, and the last 4 were numbers $0 – 9$, you'd have $26^3 \times 10^4$, since there are 3 places for 26 letters and 4 places for 10 numbers.

Square Roots

The term **SQUARE ROOT** is used to describe a number that can be squared to equal the number provided.

The radical sign ($\sqrt{\ }$) is used to show square roots. For example, $\sqrt{9} = 3$, since $3^2 = 9$.

Some numbers will have very clean whole numbers for their roots. These are known as perfect squares. Here is a table that shows common perfect squares:

Table 4.1. Perfect squares

Number	Perfect Square	Square Root
1	1	$\sqrt{1}$
2	4	$\sqrt{4}$
3	9	$\sqrt{9}$
4	16	$\sqrt{16}$
5	25	$\sqrt{25}$
6	36	$\sqrt{36}$
7	49	$\sqrt{49}$
8	64	$\sqrt{64}$
9	81	$\sqrt{81}$
10	100	$\sqrt{100}$

Note that if we wanted the number that is multiplied 3 times to get a certain number, this is the cube root, so $\sqrt[3]{8} = 2$. We can use this same notation for 4th roots, and so on.

Order of Operations

The order of operations is the way that multiple operations need to be done in order to reach the correct answer. Some operations take precedence over others; mathematical operations don't also go from left to right, like reading does.

Here is the basic order of operations:

1. First take care of any operations that are within a grouping symbol such as parentheses () or brackets [].
2. Next handle the roots and the exponents.
3. Next handle the multiplication and division in the same order that they appear (left to right).

4. Finally, handle the addition and subtraction (again, moving from left to right).

For an example of why this is important, consider the following:

Solve: $2 + 2 \times 2$

If you ignore the order of operations, what do you get?

$$2 + 2 = 4 \times 2 = 8$$

Now if you were to follow the order of operations, what happens?

$$2 \times 2 = 4 + 2 = 6$$

We get two different answers, but only the second is correct, since we need to perform multiplication before addition. It is important to eliminate any ambiguous statements in equations, because precision is key. That is why the order of operations is very important.

There is an acronym that can be used to help you remember the order of operations; **PEMDAS**:

- **P**arentheses
- **E**xponents
- **M**ultiplication/**D**ivision
- **A**ddition/**S**ubtraction

If you are using PEMDAS, you need to remember that multiplication and division have to be completed as one step from left to right, and the same with addition and subtraction.

Working with Integers

INTEGERS is a set that includes all whole numbers and the negatives (opposites) of those numbers as well. For example, integers are:

$$\ldots-3,-2,-1,0,1,2,3\ldots$$

Think of adding negative numbers as the same as subtracting positive numbers. For example,

$$6 + -5 = 6 - 5 = 1$$

ADDITION AND SUBTRACTION WITH POSITIVES AND NEGATIVES

There are two situations you will encounter when you have to add integers with negative sign(s). Are the numbers the same sign or are the numbers opposite signs? If the numbers are the same, then you can just add them like you normally would, and then add a negative sign to the answer if the two signs were negative to start out with. Here are a few example situations:

$$3 + 3 = 6$$
$$-3 + -3 = -6$$

So how do you handle a situation when the two numbers have different signs? Ignore the signs, subtract the smaller from the larger, and then use the sign of the greater of the two numbers. Here are a few examples:

$$3 + {-4} = -1 \text{: } 4 \text{ is greater than } 3, \text{ so use the sign before the } 4$$
(negative) in the answer

$$-5 + 4 = 1 \text{: } 5 \text{ is greater than } 4, \text{ so use sign before the } 5 \text{ (positive) in}$$
the answer

If you are subtracting, and have two negatives together, change them into one positive:

$$2 - {-2} = 2 + 2 = 4$$

In all, the best way to handle subtraction with negative numbers is to just turn the problem into an addition problem.

You may also need to know that the ABSOLUTE VALUE (signified by | |) is the positive equivalent of positive and negative numbers. So |−4| = 4 and |4| = 4. So a number and the absolute value of its negative counterpart are equal.

Multiplication and Division with Positives and Negatives

Again, the best way to handle this is to ignore the signs and then multiply or divide like you normally would. To figure out what sign the final product will have, you simply have to figure out how many negatives are in the numbers you worked with. If the number of negatives is even, the result will be a positive number. If the number of negatives is odd, the result will be a negative number.

For example:

$$2 \times 2 \times {-2} = -8 \text{: one negative; odd number of negatives, so}$$
negative

$$-2 \times {-2} \times 2 = 8 \text{: two negatives; even number of negatives, so}$$
positive

$$-2 \times {-2} \times {-2} = -8 \text{: three negatives; odd number of negatives, so}$$
negative

Exponents of Negative Numbers

Exponents with negative numbers are relatively simple: If the number with the exponent (the base) is negative and the exponent is even, the result will be positive. If the number with the exponent is negative and the exponent is odd, the result will be negative. This is because, for example, $-4 \times {-4} \times {-4} \times {-4} = 256$, but $-4 \times {-4} \times {-4} = -64$.

But we have to be careful with negative bases and even exponents. For even exponents, if the base is negative and in parentheses (or is a variable that you're putting in a number in for), the result is positive. For even exponents, if the negative sign is outside the parentheses, you have to raise the base to the exponent first, and then apply the negative (change the sign).

Here are some examples:

$$(-4)^2 = 16$$

$$-4^2 = -(4^2) = -16$$

$$(-3)^3 = -27$$

$$-3^3 = -(3^2) = -27$$

Evaluate x^2, where $x = -2$: $(-2)^2 = 4$

Evaluate $-x^4$, where $x = -2$: $-(-2)^4 = -16$

And again, remember that exponents are just another way of writing out multiplication.

Working with Fractions

Check the arithmetic review section of this guide for information on how to convert fractions back and forth. This section will cover different types of fractions and how to perform operations on fractions.

Equivalent Fractions

Equivalent fractions are fractions that are equal. One of the most common things that you will do when you are working with fractions is to simplify them. Another way to state this is to "reduce" fractions. All this means is writing the fraction in the smallest equivalent fraction you can (smallest top and bottom). For example, $\frac{5}{10}$ and $\frac{2}{4}$ can both be simplified to $\frac{1}{2}$. So, $\frac{5}{10}$, $\frac{2}{4}$, and $\frac{1}{2}$ are all equivalent fractions.

Here are some things to keep in mind when you are trying to simplify fractions:

- You need to find a number that can evenly divide into the top and the bottom number of the fraction that you are simplifying. After that, you can do the actual division.

- Once the division is finished, check to make sure your fraction cannot be further simplified. It is easy to make this mistake; even if you have found an equivalent fraction, it may be completely simplified, and your answer could be wrong.

- You can use simplification to reduce fractions to lower terms by dividing the top and bottom by the same number. You can also, perhaps more importantly, raise fractions to higher terms if you multiply both the top and the bottom numbers by the same number. This is very important in the addition and subtraction of fractions.

Addition and Subtraction

To add and subtract fractions, you need to understand two terms:

- **NUMERATOR:** The number on the top of a fraction.
- **DENOMINATOR:** The number on the bottom of a fraction.

If you have two numbers that have the same denominator, you will have what is known as a common denominator. You can really only add or subtract fractions that have a common denominator, so if you do not have one, you need to make one, using equivalent fractions.

Here are the steps for adding and subtracting fractions:

1. If the fractions have a common denominator, then proceed as usual. If not, then reduce or raise one or both fractions until you have a common denominator.
2. Add or subtract the numerators as you would any number, ignoring the denominator.
3. Place the resulting sum (or difference) on top of the common denominator as the new numerator.
4. Simplify the new fraction as much as possible.

Step 1 is involved with coming up with what is known as the least common denominator. This is the smallest number that all fractions have as a common denominator. The process of doing this is the same as the process for finding the least common multiple.

You can turn mixed numbers into improper fractions before adding, for example:

$$2\frac{1}{2} + 3\frac{3}{4} = \frac{5}{2} + \frac{15}{4} = \frac{10}{4} + \frac{15}{4} = \frac{25}{4} = 6\frac{1}{4}$$

You can also add the whole number parts, then add the fractional parts, and then add the two together. This will save you a lot of trouble and extra steps. It is just as correct as any other method of solving the problem and, of course, remember that you are not being tested on how you did it, just that it was done. Here is the same problem:

$$2\frac{1}{2} + 3\frac{3}{4} = 2 + \frac{1}{2} + 3 + \frac{3}{4} = 2 + 3 + \frac{1}{2} + \frac{3}{4} = 2 + 3 + \frac{2}{4} + \frac{3}{4} = 5\frac{5}{4} = 6\frac{1}{4}$$

You do have to be careful subtracting mixed numbers, since you may have to borrow:

$$2\frac{1}{4} - 1\frac{1}{2} = 2\frac{1}{4} - 1\frac{2}{4} = 1\frac{5}{4} - 1\frac{2}{4} = \frac{3}{4}$$

A simpler way may be to turn mixed fractions into improper fractions:

$$2\frac{1}{4} - 1\frac{1}{2} = \frac{9}{4} - \frac{3}{2} = \frac{9}{4} - \frac{6}{4} = \frac{3}{4}$$

Multiplication and Division

Multiplication and division of fractions is actually simpler than adding or subtracting them. When you have to multiply fractions together, you'll first want to turn any mixed fractions into improper fractions. Then you just multiply the numerators to get the new numerator and then multiply the denominators to get the new denominator. Once that is done, you can go ahead and simplify. For example:

Inverting a fraction changes multiplication to division:
$$\frac{a}{b} \div \frac{c}{d} = \frac{a}{b} \times \frac{d}{c} = \frac{ad}{bc}$$

$$2\frac{1}{5} \times \frac{3}{7} = \frac{11}{5} \times \frac{3}{7} = \frac{11 \times 3}{5 \times 7} = \frac{33}{35}$$

There is no way to simplify $\frac{33}{35}$, so this is the final answer.

Division is a little bit more complicated, but ultimately it is the same procedure. First, you have to find the reciprocal of the second fraction. A reciprocal is a fraction that is "flipped": its numerator and denominator are switched. Once that is done, you will multiply the fractions as usual. Here is an example:

$$\frac{5}{6} \div \frac{2}{3} = \frac{5}{6} \times \frac{3}{2} = \frac{5 \times 3}{6 \times 2} = \frac{15}{12} = \frac{5}{4}$$

As you can see, the second fraction $\left(\frac{2}{3}\right)$ simply flips to $\frac{3}{2}$. The common way this is explained is to "flip and multiply". That is as good an explanation as any, and is certainly easier to remember. Again, you are being tested on your ability to do the math here, not to know the jargon.

Algebra

ALGEBRA is a method of generalizing expressions involved in arithmetic. You will be able to explain how groups of things are handled all the time. This is useful for times when you have a certain function that you need to do over and over again. In algebra, you typically have numbers as well as symbols (usually letters) called VARIABLES that can be used to stand for certain numbers. Typically, variables are useful when dealing with word problems, since algebra makes it easier to solve for an "unknown."

Evaluating Numbers

Numbers that are assigned a definite value (like the number "1") are constants. When symbols (variables) are used to stand for numbers, they can typically take on any number. If you saw the equation: $2x = 6$, x would be the variable here, and in this case it would equal 3.

A few things to keep in mind:

- You can make a variable anything you want. X, y, Z, a, A, b, and so on are examples. They are usually italicized, to distinguish them from numbers.
- Both sides of the = sign are, obviously, =. So you can add or subtract or multiply or divide anything on both sides (constants or variables or both), as long as you do it to both sides. This is how you solve equations. With inequalities, however, you must change the sign if you multiply or divide by a negative number.
- Many times, you must distribute either variables or constants through to get rid of parentheses. For example, if you have $3(x - 9)$, you'll want "push through" the 3 to make it $3x - 27$.
- Once you have figured out what a variable is, plug it into the original equation to make sure everything is still equal.
- You may be asked to evaluate a function for a certain value in the variable. Just plug in that value everywhere the variable is. For example, the value of $f(3)$ in the function $f(x) = 3x + 2$ is $3(3) + 2 = 11$.

Equations

Equations are expressions that have an equal sign, such as $2 + 2 = 4$. Equations will usually have a variable, and will always be true or false. $2 + 2 = 1$ is false. $2 + 2 = 4$ is true. When you are solving for variables, only one answer for each variable, usually, will make the expression true. That is the number you must find.

Basically, to do this, you will just rewrite the equation in more and more simple terms until you have the solution to it. Ideally, you want this to be x (or whatever variable) = a number.

$$x = \#$$

Again, you can do anything to an equation as long as you do the same thing to both sides. This is how you solve equations.

EXAMPLE

Solve $3(x - 2) = -7 + 10$.

First, simplify anything you can on one side (the –7 and 10) and also get rid of any parentheses by "pushing through" (the "3" on the left hand side):

$3x - 6 = 3$

Next, add 6 to both sides to get $3x$ by itself:

$3x - 6 + 6 = 3 + 6$

$3x = 9$

Note that if we have an inequality instead of an equation, we have to be careful. We solve an inequality the same way, but if we multiply or divide by a negative number, we have to switch the sign.

EXAMPLE

Solve $-3x - 2 \geq 4$.

$-3x - 2 \geq 4$

$-3x - 2 + 2 \geq 4 + 2$

$-3x \geq 6$

$\frac{-3x}{-3} \leq \frac{6}{-3}$

$x \leq -2$

One more thing to mention here: If you are given an algebraic equation that is has 0 on one side and factors with variables, set each factor to 0 to solve for the variable. For example:

Solve for x: $(2x + 6)(3x - 15) = 0$

$2x + 6 = 0$ and $3x - 15 = 0$; $x = -3, 5$

An algebraic function like the one above is a polynomial of degree 2, which is a quadratic. To get the degree of a polynomial, add up all the exponents in each term, and the degree is the sum of exponents in the highest term. For example, for $4x^3y^5 + 8x^8y + 4x + 5$, the degree is 9, since in the second term, which has the largest sum of exponents, the sum of the exponents is $8 + 1 = 9$.

Word Problems

WORD PROBLEMS often utilize algebraic principles in their text. It is important to know how to properly assign a variable inside of a word problem. Pay attention to the words used: x *equals* or *a number equals*, x *is less than*, *2 is added to* x, are common wordings. A lot of times, you can just "translate" the English right into the math; for example, *a number added to 5* would be $x + 5$. Be careful though; *3 less than a number* would be $x - 3$ (do the math with simple numbers to figure out wording).

If you are not given a specific variable but you need one, just call it whatever you want. x is the simplest variable you can use to do this, and is also one of the least confusing when you start working with more complex algebraic principles.

Here is a type of word problem you might encounter with ratios: A class of 140 has sophomores, juniors, and seniors in a ratio of 4:2:1. How many juniors are in the class? To solve this, you can set it up like this: $4x + 2x + 1x = 140$; $7x = 140$; $x = 20$. Since there are $2x$ juniors, there are $2 \times 20 = 40$ juniors in the class.

There is another type of algebra problem that you might encounter, and it involves "work". You can remember this equation:

$$\frac{\text{Time to do a job together}}{\text{Time to do a job alone}} + \frac{\text{Time to do a job together}}{\text{Time to do a job alone}} = 1.$$

For example, if we had a problem like "If Mark can do a job in 4 hours and John can do it in 3 hours, how long would it take for the two to do the job working together?," we'd have $\frac{x}{4} + \frac{x}{3} = 1$, to get $x = \frac{12}{7}$.

Also, always remember that Distance = Rate × Time.

Multiplying Exponents

Multiplying variables with exponents is simple, as long as you have the same number with the exponent (the "base"). Just keep the same base and add the exponents together to get the new exponent. Keep in mind how exponents work as well; that is important.

Here is an example of exponents with variables:

$$x \times x \times x = x^3$$

Here is an example showing the multiplication of exponents; you can see why you add exponents:

$$x^2 \times x^3 = x^{2+3} = x^5, \text{ since } (x \times x) \times (x \times x \times x) = x^5$$

When raising something with an exponent in it to another exponent, you multiply exponents. And when dividing exponents, you do the opposite of addition, and subtract exponents with the same base. You may get a negative exponent; if you do, you can put it on the other side of the "division sign" and make it positive. For example:

$$\frac{(2x^2y)^3}{x^4y^5} = \frac{8x^6y^3}{x^4y^5} = 8x^{(6-4)}y^{(3-5)} = 8x^2y^{-2} = \frac{8x^2}{y^2}$$

Polynomials

POLYNOMIALS are just a list of algebraic terms that can contain variables, constants, and exponents, but can never have any term that is divided by a variable.

To combine like terms, you can add or subtract terms with the exact same variables. For example: $3x^2 + 2xy - 4 + 4x^2 - xy = 7x^2 + xy - 4$.

Note that if you have to multiply two two-term expressions to get a polynomial, use the FOIL (First, Outer, Inner, Last) method. For example:

$$(3x + 2)(x - 2) = (3x)(x) + (3x)(-2) + 2x - 4 = 3x^2 - 6x + 2x - 4 = 3x^2 - 4x - 4$$

If you have the same two-term expressions but with opposite signs (difference of squares), the middle terms (Outer and Inner) will cancel out:

$$(x + 2)(x - 2) = (x)(x) + (x)(-2) + 2x - 4 = x^2 - 4$$

Factoring

FACTORING is the process of breaking one quantity down into the product of some other quantity (or quantities). When you learned to use distribution to multiply out variables ("push through"), you are learning to do the opposite of factoring. When you factor, you are removing parts of an expression in order to turn it into factors. Basically, you do this by removing the single largest common single term (monomial) that is a factor; this is the greatest common factor.

> ## EXAMPLE
>
> Factor the following: $3x^2 - 9x$
>
> Basically, you will pull out the largest common denominator between the two of them. In this case, it is $3x$. So you will be left with the following:
>
> **$3x (x - 3)$**

To check this, "push through" the $3x$ to both the x and the 3, with a "−" sign in the middle; you get the original! It is also possible to do this with expressions that have more than two terms. Usually, these will be trinomials, with 3 terms. These will always end up being in the form of (x plus or minus something) times (x plus or minus something). What the second number is and what the sign is for the problem is determined by the original problem (we saw this above in the Exponents – Multiplication section).

Also, if you have a difference of squares, as we saw above, you can factor as the following:

$$(a^2 - b^2) = (a - b)(a - b)$$

You can SIMPLIFY algebraic expressions the exact same way that you would fractions. Basically, you cake out the common factors. They just cancel out. You will want to multiply everything out if you cannot find a factor and then simplify. Here is an example:

$$\frac{(x + 4)^2}{x + 4} = \frac{(x + 4)(x + 4)}{x + 4} = x + 4$$

The other thing to remember is if you have to add fractions with variables, you may have to find a common denominator. For example:

$$\frac{x - 2}{x} + \frac{x - 2}{x + 2} = \frac{(x - 2)(x + 2)}{x(x + 2)} + \frac{x(x - 2)}{x(x + 2)} = \frac{x^2 - 4 + x^2 - 2x}{x(x + 2)} = \frac{(2x^2 - 2x - 4)}{x(x + 2)}$$

Geometry

The GEOMETRY section here is pretty straightforward. See the arithmetic reasoning portion of the guide for examples of perimeter and area. New information will be covered here.

Angles

ANGLES are typically measured in degrees. As an example, if you were to completely rotate around a circle a single time, you would have gone 360 degrees, since there are 360 degrees in a circle. Half a circle is 180 degrees, a quarter circle is 90 degrees, and so on. Degrees are used to talk about what fraction of a total rotation around a circle a certain angle represents.

Table 4.2. Types of angles

TYPE OF ANGLE	DEFINITION
Acute angles	Angles with a measurement of fewer than 90 degrees
Right angles	Angles with a measurement of exactly 90 degrees
Obtuse angles	Angles with a measurement of more than 90 degrees
Complementary angles	Two angles which add up to 90 degrees
Supplementary angles	Two angles which add up to 180 degrees

Note that the sum of two angles that form a straight edge is 180 degrees. Note also that angles corresponding angles of parallel lines are equal.

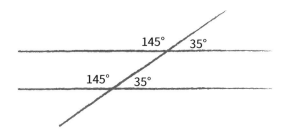

Figure 4.1. Corresponding angles of parallel lines

Triangles

TRIANGLES are geometric figures that have three sides (which are straight). The most important thing you need to know about triangles is that the sum of the measurements of the angles always equals 180. This is important because it means that if you know two of them, you can very easily calculate the third simply by subtracting the first two from 180. There are three types of triangles that you need to know about.

Table 4.3. Types of triangles

Type of Triangle	Definition
Equilateral triangles	These are triangles in which the three angles are the same measure. Each one of the angles is 60 degrees.
Isosceles triangles	These are triangles which have two sides of the same length. The two angles which are directly opposite the sides of the same length will be the same angle. If B was the third angle, then you could state that lines between A and B (AB) and between lines B and C (BC) are the same. $AB = BC$.
Right triangles	These might be the most important. A right triangle is a triangle which has one side that equals 90 degrees. Two sides are legs and the third side, which is directly opposite the 90-degree angle, is the hypotenuse, the longest one of the sides.

With right triangles, you can always remember that the lengths of the sides are related by the following equation (the Pythagorean Theorem, for those of you who remember):

$$a^2 + b^2 = c^2$$

Remember that c is the hypotenuse, which is the side opposite the right angle. It is the longest of the sides.

The area of triangles is $\frac{bh}{2}$, where the base is perpendicular (form a right angle) to the base.

Also, with right triangle with two legs that are equal, the hypotenuse is always $\sqrt{2}$ times the value of a smaller side. For right triangles with 30-60-90 angles, the hypotenuse is always twice the smaller side (side across from the 30 degree angle), and the other leg (across from the 60 degree angle) is $\sqrt{3}$ times the smaller side.

Circles

A CIRCLE, as you probably are already aware, is a closed curve where every single point is the exact same distance from its center. If you draw a line from the outside of the circle to the fixed point in the center, you have the radius. If you draw a line from one side of the circle straight through to the other, passing through the center point, you have the diameter. A chord is a line from one point of the circle to another, but not necessarily a diameter. By the definition of a circle, all radii (plural of radius) are equal and all diameters are equal.

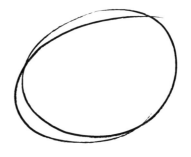

Figure 4.2. Diameter and radius

For a circle, remember that diameter = 2 × radius.

You can use the radius to find out the circumference of a circle (how big it is around), by using the following formula, where C = circumference and r = radius:

$$C = 2\pi r$$

Think of the circumference like the perimeter of the circle; if you were to take a piece of string all the way around a circle and measure it, this would be the circumference.

To find the area of a circle, use the following formula, where A = area and r = radius.

$$A = \pi r^2$$

The area is the measurement of what's inside the circle and is written in square units.

The Coordinate System

The coordinate system or, sometimes, the Cartesian coordinate system, is a method of locating and describing points on a two-dimensional plane; It is a reference system. The plane is two number lines that have been laid out perpendicular to each other, with the point that they cross being origin (0,0). The origin is the point 0 for both the x (horizontal) and y (vertical) axes. Positive and negative integers are both represented in this system.

In the Figure 8.3, each small tick on the line is equal to 1. The larger ticks represent multiples of 5. A point is also depicted, P, which shows how things are placed onto the coordinate system. Again, the horizontal line on the coordinate plane is called the x-axis, and the vertical line on the coordinate plane called the y-axis. The points, when described, are described in reference to where they lie on that plane.

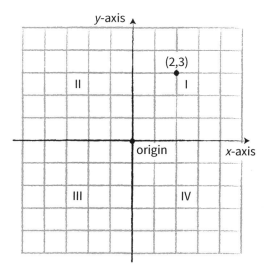

Figure 4.3. The coordinate system

Here is an example that is the point shown above:

$$(2,3)$$

The first number, 2, is the x-axis. The second, 3, is the y-axis. So this point is at position 2 of the x-axis (go over 2 units to the right) and position 3 of the y-axis (go up 3 units). Note that for negative x values, we go to the left that many units, and for negative y values, we go down.

A way to simplify the way these points work is to say:

$$(x,y)$$

Slope

The SLOPE is the steepness of a given line. When you look on the coordinate plane, if you draw a line between two points, you will get the slope. One of the main uses of a slope is to signify a rate; it measures how much a certain thing goes up for every unit it goes over.

Here is the way to find slope:

$$\text{Point A: } (x_1, y_1)$$
$$\text{Point B: } (x_2, y_2)$$
$$\text{Slope} = \frac{y_2 - y_1}{x_2 - x_1}$$

In plain English, you will take the difference between the y-coordinates and divide them by the difference between the x-coordinates. Remember: y is vertical and x is horizontal.

The following is an example of how this type of problem would likely manifest:

Some things to remember:

- If the slope is positive, the line is going up as it goes right.
- If the slope is negative, the line is going down as it goes right.
- The larger the absolute value (positive value) of the slope, the steeper the line.
- If a line is horizontal, its slope is 0; if a line is vertical, its slope is undefined.
- The formula for a line is $y = mx + b$, where m is the slope and b is the y-intercept (where the line crosses the y-axis).
- If two lines are parallel, they have the same slope. If they are perpendicular, they have negative reciprocal slopes.
- If two lines are intersection, the x and y values work in both equations; in other words, if you were to plug in the x and y values, both equations would be true. This is how you "solve" a system of equations.

Tips

Here are some tips to help you make it through the mathematics knowledge section of the AFOQT:

- Remember to utilize PEMDAS and the order of operations when you are working through problems.
- The more you practice, the easier the problems will be. There are a lot of internet sites with math problems.
- Be careful with the answers. The test will often provide common mistakes answers among the correct answer.
- Since you will not have access to a calculator, you will need to round π, 3.14, or $\frac{22}{7}$ are the most common ways to do this. Sometimes you may be asked to simply include it in your answer without actually utilizing the digits in your calculations at all.

- Don't mix up the perimeter and area formulas.

- One way to get good at algebra relatively quickly is to set up all of your problems, even simple ones, as algebra problems. Remember, you can simply create a variable and stick it on one side of the equation to solve for it. For example, *3 times a number added to 10 is 16* becomes $3x + 10 = 16$. Translate almost word-for-word from English to math, and then get *x* by subtracting 10 from each side, and then diving by 3.

- Keep in mind that, unlike many tests you probably remember taking in high school, you are not being tested on your ability to write out how you solved the problem. There are many correct ways to solve mathematical problems and, as long as you come to the right answer in the end, it does not matter how you solved it. Nobody will be coming behind you and checking your scratch paper to see what you did (unless they think you are cheating, but that is a whole other ball of wax).

- If you can't solve the problem directly, trying plugging in the answers to see if you can figure out which one works.

- Pay close attention to the positive and negative signs in the work that you are doing. This is extremely important because, again, they will likely throw the correct answers out there as one of the choices but with the wrong sign.

- Always make sure the fractions that you are working with are simplified when you finish with them. Unless otherwise stated, the test answers will usually want the most simplified version of the fraction.

- Make sure you go step by step through each question. Don't skip steps or combine steps. Doing either could lead to an issue where something is accidentally missed.

- It might help you to change even the normal expressions into algebra. Often, understanding what you are being asked to do and knowing how to handle certain problems is only made easier when you are using algebra to handle it.

- Draw pictures on geometry problems.

GO ON

Practice Questions

1. Evaluate the expression $\frac{4x}{x-1}$ when $x = 5$.

   1-A 3

   1-B 4

   1-C 5

   1-D 6

2. Simplify: $3x^3 + 4x - (2x + 5y) + y$

   2-A $3x^3 + 2x + y$

   2-B $11x - 4y$

   2-C $3x^3 + 2x - 4y$

   2-D $29x - 4y$

3. Simplify: $(x + 7)(x - 5)$

   3-A $x^2 + 2x - 35$

   3-B $x^2 \pm 2x - 35$

   3-C $35x$

   3-D $7x^2 + 5x - 35$

4. Simplify the expression $\frac{4xy^3}{x^5y}$.

   4-A $\frac{12}{x^4}$

   4-B $12(x^2y)^2$

   4-C $64(x^2y)^2$

   4-D $\frac{64y^2}{x^2}$

5. Find the area of a rectangular athletic field that is 100 meters long and 45 meters wide.

   5-A 290 m

   5-B 4,500 m²

   5-C 145 m²

   5-D 4.5 km²

6. Mary runs 3 miles north, 4 miles east, 5 miles south, and 2 miles west. What are her final coordinates (in miles), with respect to her starting point?

   6-A $(8, 6)$

   6-B $(-2, 6)$

   6-C $(7, 3)$

   6-D $(2, -2)$

7. Two identical circles are drawn next to each other with their sides just touching; both circles are enclosed in a rectangle whose sides are tangent to the circles. If each circle's radius is 2 cm, find the area of the rectangle.

   7-A 24 cm²

   7-B 8 cm²

   7-C 32 cm²

   7-D 16 cm²

8. Solve for a: $3a + 4 = 2a$

   8-A $a = -4$

   8-B $a = 4$

   8-C $a = \frac{-4}{5}$

   8-D $a = \frac{4}{5}$

9. Evaluate the expression $|3x - y| + |2y - x|$ if $x = -4$ and $y = -1$.

   9-A -11

   9-B 11

   9-C 13

   9-D -13

10. Solve for x: $8x - 6 = 3x + 24$

    10-A $x = 3.6$

    10-B $x = 5$

    10-C $x = 6$

    10-D $x = 2.5$

11. Convert 0.25 into a percentage.

    11-A 250%

    11-B 2.5%

    11-C 25%

    11-D 0.25%

12. $-2x - 3x^2 + 4x =$

    12-A $-3x^2 - 2x$

    12-B $-3x^2 + 2x$

    12-C $3x^2 + 2x$

    12-D $3x^2 - 2x$

13. Convert $\frac{38}{98}$ into a percentage. Round to the nearest percent.

    13-A   39%

    13-B   38%

    13-C   38.77%

    13-D   38.775%

14. Factor the following: $x^2 + 4x + 4$

    14-A $2(x + 2)(x + 2)$

    14-B $(x - 2)(x - 2)$

    14-C $(x - 2)(x + 2)$

    14-D $(x + 2)(x + 2)$

15. Solve for x: $4x + 3 > -9$

    15-A $x < 3$

    15-B $x > -3$

    15-C $x > -1\frac{1}{2}$

    15-D $x < 1\frac{1}{2}$

GO ON

Math Knowledge
Answer Key

1.	C.	9.	C.
2.	C.	10.	C.
3.	A.	11.	C.
4.	D.	12.	B.
5.	B.	13.	A.
6.	D.	14.	D.
7.	C.	15.	B.
8.	A.		

five

INSTRUMENT COMPREHENSION

The Instrument Comprehension section will test your ability to understand basic aviation instrumentation by providing two dials, a compass, and artificial horizon. Then it requires you to identify an image of an aircraft that correlates to what is shown on the dials. It might seem difficult since you may not have any previous flight experience, but once you understand the dials, most find this section to be relatively easy and straightforward.

Let's review the dials first, and then we'll explain how the questions are formatted and how you need to use the dials to answer the questions. In each question, you will be presented two dials that look just like this:

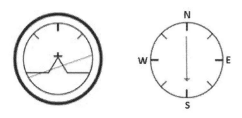

The dial on the left is your artificial horizon. This dial can be confusing the first time you see it. In the artificial horizon pictured above, the dial indicates that the aircraft is in a slight climb and banking right. What makes this confusing to people, is that it is the *horizon* that moves on the dial.

The easiest way to explain it is to imagine it is the view from the aircraft. Here's how it works: the line with a triangle in the middle with the "plus sign" is intended to be the wings and nose of the aircraft. Now, think of that in relation to the diagonal line cutting across the dial, sloping to downward to the left. That sloping line is the horizon. As the aircraft climbs upward, the horizon will dip below level, just like it would be viewed from the cockpit. Then, as the aircraft banks right, visualize in your mind the wings of the aircraft turning in relation to the horizon. After you try a few, it becomes easier and easier to immediately visualize.

The dial on the right likely needs no explanation. It is a simple compass indicating the direction of travel of the aircraft. We will come back to this feature later, as it is a great way to quickly and easily eliminate wrong answers.

Now that you've been introduced to both dials, let's review the artificial horizon by going over some examples.

In this example, the horizon is flat and level at the center point. That means there is no climb or dive, and the aircraft's wings are flat and level as represented by the image of the plane.

In the above artificial horizon, the aircraft is in a dive. As you can see, the horizon has moved above the fuselage silhouette (the line with the triangle representing the aircraft). Try to imagine you are looking through the windshield of a plane as it starts to dive. As the nose tips down, the horizon will move upward in your field of vision. That's exactly what the artificial horizon represents.

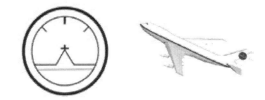

Similar to an aircraft diving, is if it is climbing. Again, imagine sitting in the cockpit as the plane climbs. In your field of vision, the horizon would go down, down, down as the nose turns up. Again, that's exactly what the artificial horizon is doing in the dial.

Banking is where most people get confused. Most people who experience frustration with this are typically visualizing the angle of the horizon as the angle of the wings. This happens because the dial is from the perspective of the pilot in the aircraft. However, on the AFOQT exam, the images you are shown are from outside the aircraft and could be viewed from the front, back, or sides. Therein lies the challenge, to imagine the perspective of the pilot and translate that into the image you are shown. As you can see in the above example, the horizon is sloping down to the right, but the wings of the plane are turning leftward as it banks left. Again, visualize sitting in the aircraft as it banks and how the horizon would appear in relation to the windshield.

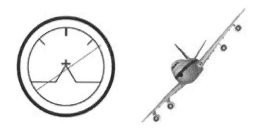

Here is a final example that throws you one last curveball. In this example, the aircraft appears to be coming towards you. If the aircraft is coming towards you, the artificial horizon will appear to tilt the same way as the wings because your perspective is mirrored to the direction of travel. If you get confused, remember first to think about what direction the aircraft is banking (to its own right or it's own left), then correlate that to the image of the aircraft.

Now, let's put it all together in an example of an actual test question. The questions will appear on the exam much like the one on the following page.

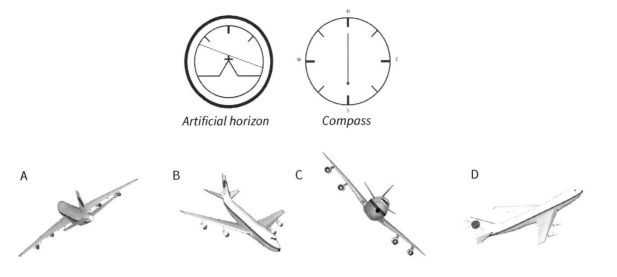

Artificial horizon   *Compass*

A   B   C   D

As we mentioned earlier, we are coming back to the compass. This is the easiest way to eliminate wrong answer choices, and sometimes immediately find the correct answer. You always want to reference the artificial horizon just to double check (sometimes options on the test are very similar), but it is fastest and easiest to start with the compass. In this example, we see the aircraft is heading south. On the exam, the direction of travel is simply that *south* is coming towards you, *north* is flying away, *east* is to the right, and *west* is to the left. Almost too easy, right? In this example, the compass indicates *south* and looking at the choices, there is only one aircraft heading directly towards you, which is choice C. To double check, the artificial horizon dial indicates that the aircraft is neither diving nor climbing but is banking relatively hard to the left. Choice C confirms that, as that aircraft is shown to be neither climbing nor diving and is indeed banking hard to the left. The answer is C.

Now that you've got a decent grasp of the fundamentals and have seen how the parts of the puzzle fit together, it's time to get some practice in. Keep in mind that on the actual test, you will have six minutes to answer twenty questions. That's about twenty seconds per question which might seem fast, but once you have some practice under your belt, you'll be able to recognize the correct answer in just a few seconds. You will have ample time to double check all the answer choices before moving on to the next question. Always look at all of the answer choices before selecting your answer!

Practice Problems

1.

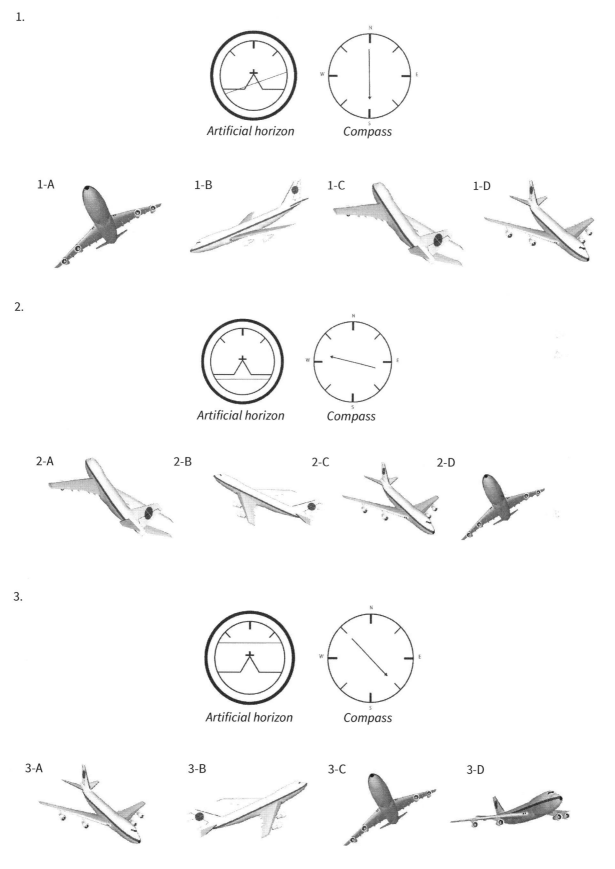

Artificial horizon   Compass

1-A   1-B   1-C   1-D

2.

Artificial horizon   Compass

2-A   2-B   2-C   2-D

3.

Artificial horizon   Compass

3-A   3-B   3-C   3-D

4.

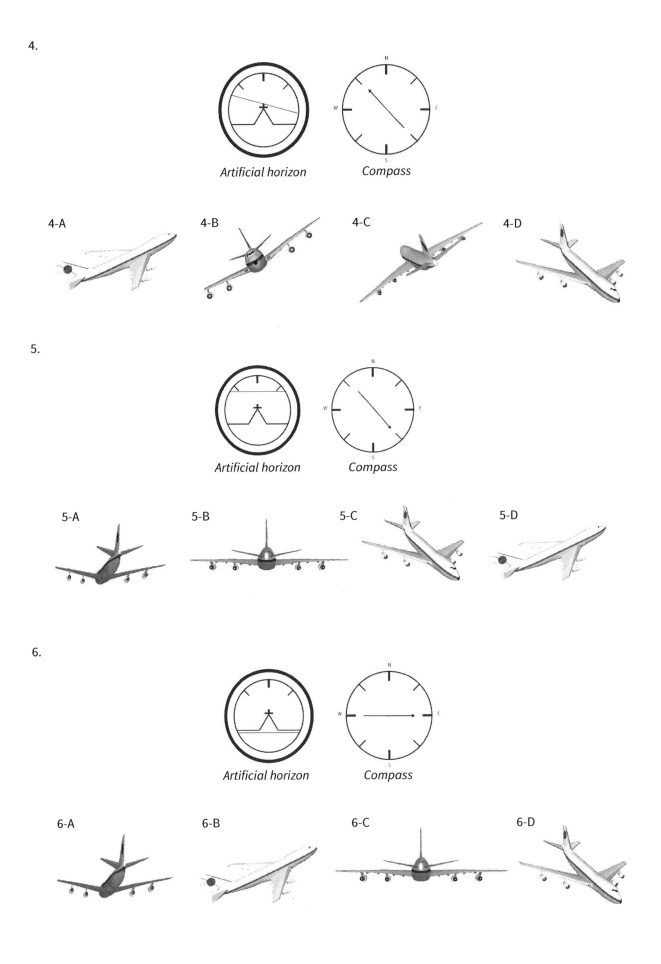

Artificial horizon   Compass

4-A   4-B   4-C   4-D

5.

Artificial horizon   Compass

5-A   5-B   5-C   5-D

6.

Artificial horizon   Compass

6-A   6-B   6-C   6-D

7.

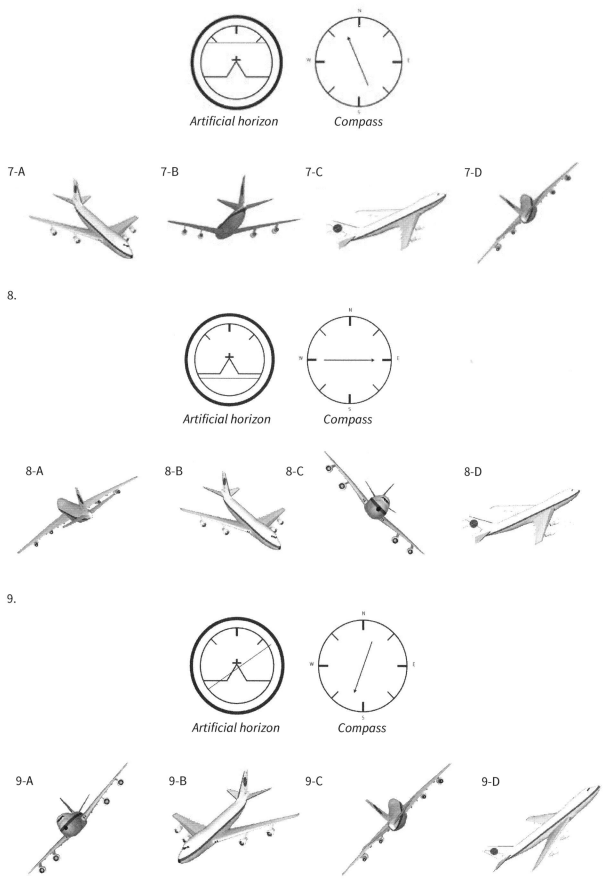

Artificial horizon   Compass

7-A   7-B   7-C   7-D

8.

Artificial horizon   Compass

8-A   8-B   8-C   8-D

9.

Artificial horizon   Compass

9-A   9-B   9-C   9-D

10.

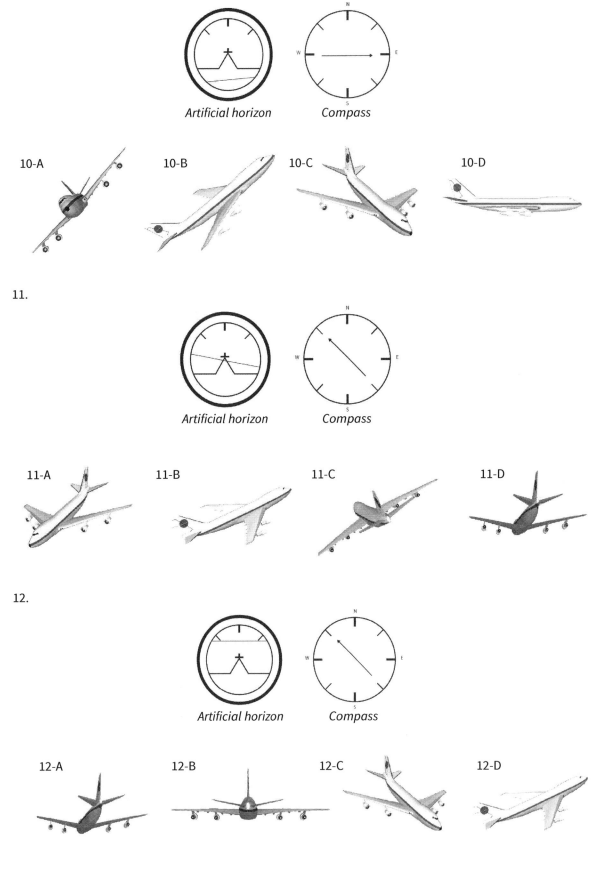

Artificial horizon   *Compass*

10-A   10-B   10-C   10-D

11.

Artificial horizon   *Compass*

11-A   11-B   11-C   11-D

12.

Artificial horizon   *Compass*

12-A   12-B   12-C   12-D

13.

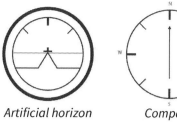 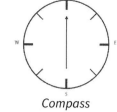

Artificial horizon   Compass

13-A

13-B

13-C

13-D

14.

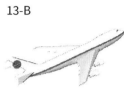 

Artificial horizon   Compass

14-A

14-B

14-C

14-D

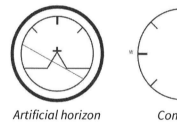 

15.

Artificial horizon   Compass

15-A

15-B

15-C

15-D

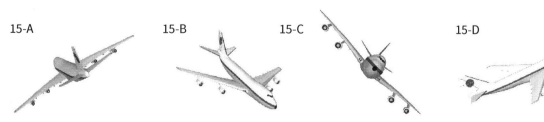 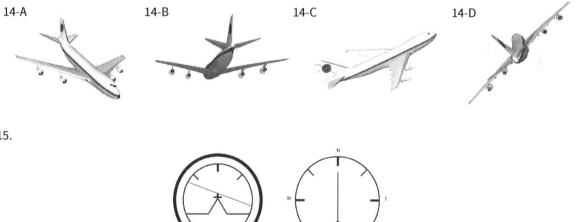

16.

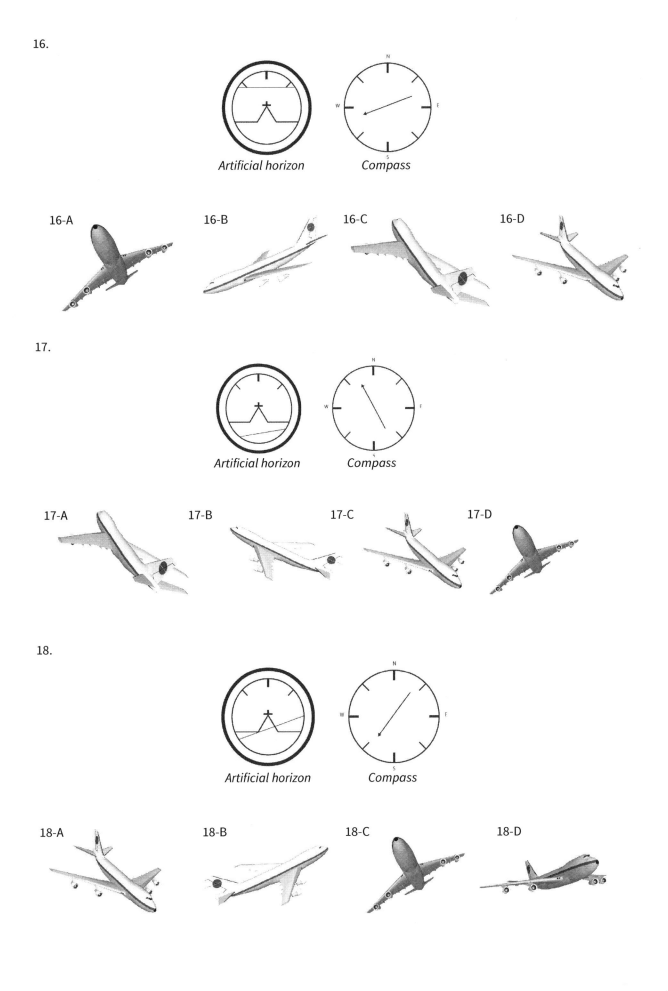

16-A   16-B   16-C   16-D

17.

17-A   17-B   17-C   17-D

18.

18-A   18-B   18-C   18-D

19.

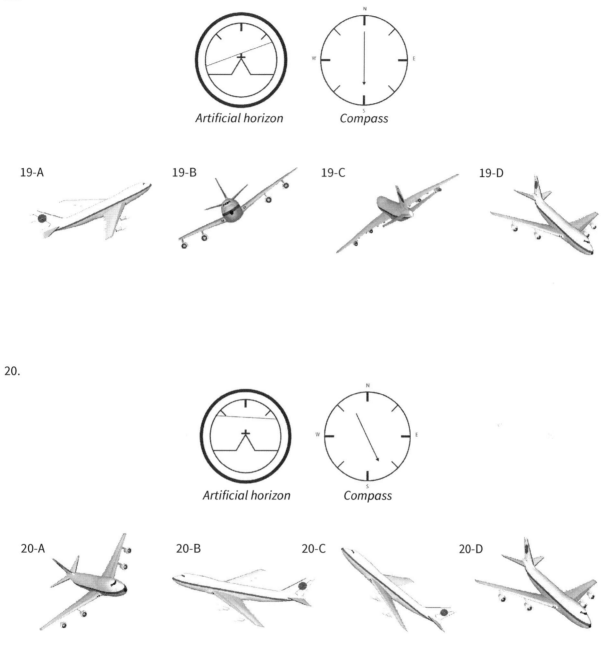

19-A   19-B   19-C   19-D

20.

Artificial horizon   Compass

20-A   20-B   20-C   20-D

GO ON

Instrument Comprehension

Answer Key

1.	A	11.	C
2.	B	12.	A
3.	A	13.	C
4.	C	14.	D
5.	C	15.	C
6.	B	16.	B
7.	B	17.	A
8.	D	18.	C
9.	A	19.	B
10.	B	20.	D

BLOCK COUNTING

The Block Counting section is relatively straightforward, but can be challenging for some people. You will be shown a shape with different stacked blocks. Some of the blocks will have a number on them, and your objective is to determine how many other blocks touch the block indicated. The challenge is that you must visualize three-dimensionally how these blocks move through the entire structure. Most mistakes come from counting a block twice on accident.

Let's look at an example. In the below image, you can see five numbered blocks. If you were asked to count how many blocks touch block 3, what would your answer be?

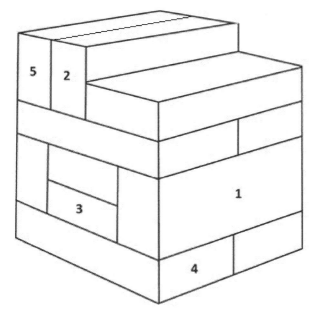

The correct answer is five however, some people will have landed on six. Why is this? People who have difficulty visualizing three-dimensional images might have counted block 4 twice because they will count blocks on one side of the shape that run past the block in the question, then count blocks from the other side of the shape they can see. The block below is the same as the one in the example, but with the blocks on one side shaded.

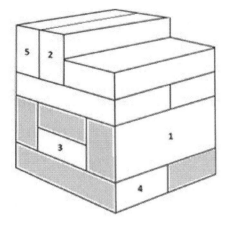

The mistake is counting the other side of block 4 as shown below with the hash marks. You have to keep in mind which blocks you have already counted and try to visualize the shapes moving past each other and how they connect.

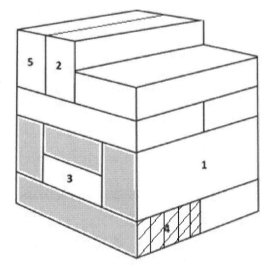

While this section is straightforward, it can be challenging. To make it even more so, keep in mind, you have only three minutes to complete twenty questions. That is a mere nine seconds per question! Practice now and see how you do. If you struggle with this section, don't forget that accuracy is just as important as speed. It does no good to answer twenty questions incorrectly when you could have answered ten correctly.

If you struggle with this section, the only way to get better is to practice, practice, practice. Use the example questions and make your own numbers for individual blocks and practice counting around them to sharpen your skills, and then go back and work on speed on the practice test.

Practice Problems

Use the block below for questions 1 – 4.

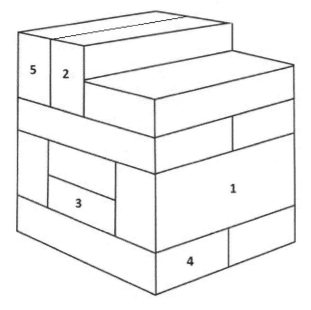

4. Block 5 is touched by _____ other blocks?

    4-A   2

    4-B   3

    4-C   4

    4-D   5

    4-E   6

Use the block below for questions 5 – 8.

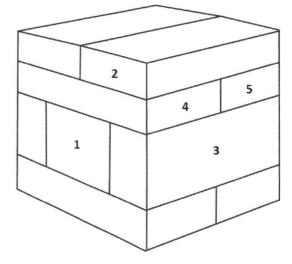

1. Block 1 is touched by _____ other blocks?

    1-A   2

    1-B   4

    1-C   6

    1-D   7

    1-E   8

2. Block 2 is touched by _____ other blocks?

    2-A   2

    2-B   3

    2-C   4

    2-D   5

    2-E   6

3. Block 4 is touched by _____ other blocks?

    3-A   3

    3-B   5

    3-C   6

    3-D   4

    3-E   2

5. Block 1 is touched by _____ other blocks?

    5-A   2

    5-B   4

    5-C   6

    5-D   7

    5-E   8

6. Block 2 is touched by _____ other blocks?

    6-A   2

    6-B   3

    6-C   4

    6-D   5

    6-E   6

7. Block 4 is touched by _____ other blocks?

   7-A   3

   7-B   5

   7-C   6

   7-D   4

   7-E   2

8. Block 3 is touched by _____ other blocks?

   8-A   2

   8-B   3

   8-C   4

   8-D   5

   8-E   6

11. Block 3 is touched by _____ other blocks?

    11-A   6

    11-B   5

    11-C   3

    11-D   4

    11-E   2

12. Block 5 is touched by _____ other blocks?

    12-A   7

    12-B   6

    12-C   4

    12-D   5

    12-E   8

Use the block below for questions 9 – 12.

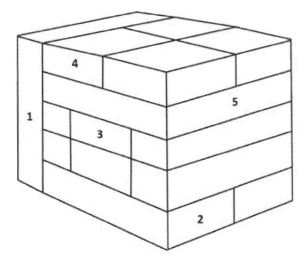

Use the block below for questions 13 – 16.

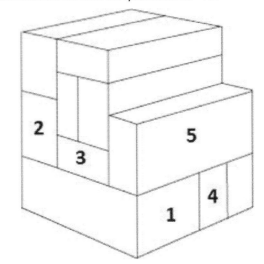

9. Block 1 is touched by _____ other blocks?

   9-A   2

   9-B   4

   9-C   6

   9-D   7

   9-E   8

10. Block 2 is touched by _____ other blocks?

    10-A   2

    10-B   3

    10-C   4

    10-D   6

    10-E   5

13. Block 1 is touched by _____ other blocks?

    13-A   2

    13-B   4

    13-C   6

    13-D   7

    13-E   8

14. Block 2 is touched by _____ other blocks?

    14-A   2

    14-B   3

    14-C   4

    14-D   6

    14-E   5

15. Block 3 is touched by _____ other blocks?

    15-A   7

    15-B   5

    15-C   3

    15-D   4

    15-E   6

16. Block 5 is touched by _____ other blocks?

    16-A   7

    16-B   6

    16-C   4

    16-D   5

    16-E   8

Use the block below for questions 17 – 20.

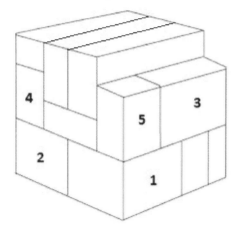

17. Block 1 is touched by _____ other blocks?

    17-A   5

    17-B   4

    17-C   6

    17-D   7

    17-E   8

18. Block 2 is touched by _____ other blocks?

    18-A   2

    18-B   5

    18-C   4

    18-D   6

    18-E   7

19. Block 3 is touched by _____ other blocks?

    19-A   6

    19-B   5

    19-C   3

    19-D   4

    19-E   2

20. Block 5 is touched by _____ other blocks?

    20-A   7

    20-B   6

    20-C   4

    20-D   5

    20-E   8

Block Counting
Answer Key

1.	C	11.	A
2.	C	12.	E
3.	D	13.	B
4.	B	14.	D
5.	C	15.	A
6.	B	16.	D
7.	C	17.	A
8.	D	18.	B
9.	D	19.	A
10.	E	20.	C

seven

TABLE READING

In the Table Reading section, the objective is simple enough. For each question, you are given X and Y coordinates. Find that intersection on the table and identify the number located there. Of all the sections on the AFOQT where time is a factor, none compares to Table Reading. You will only have seven minutes to complete forty questions, an absurdly short amount of time at only ten and one-half seconds per question.

The general concept is easy enough, any sixth grader could understand how to read a simple chart. So how do you prepare for this section to overcome the obstacle of time? Practice is, of course, important, but using a systematic approach is imperative. Have you ever tried counting backward from one hundred to zero while someone is yelling random numbers in your ear? The Table Reading section is similar to that. You will see so many numbers, it becomes extraordinarily easy to forget what the coordinates were you needed and have to go back and reference the question again. By doing that, you would have already wasted five to ten seconds minimum, which is one more question you now do not have time to answer.

Your goal is to not worry about time at all until the last thirty seconds. As you practice, force yourself to focus ONLY on the question at hand. Do not think about how much time has gone by, how many questions you have left, if you made a mistake on the last question, etc. The real test of this section is to see if, as an officer in the US Air Force, you possess the capability to stay calm under pressure, maintain your bearing, and stay focused.

As mentioned, don't worry about time until the last thirty seconds or so. When that time hits, go ahead and mark an answer for all remaining questions. There is no penalty for wrong answers on the AFOQT, so you should never, ever leave a question blank on any section of this exam.

Let's get started on practice, so you can get a feel for this section. Do not overwhelm yourself with this section. It is not possible to substantially speed up your time any more than after the first ten to fifteen practice questions you try. It is simply a function of time and it takes more than ten and one-half seconds to read the question, scroll through the table, find the answer, and mark your answer choice. Rushing is counter-productive. Your goal with this practice round is just to familiarize yourself with the format, and then move on. Do NOT take this practice test untimed! You want to know approximately how many questions you can answer in seven minutes normally so that you can better track your progress on test day and not lose track of time.

For purposes of this practice test, you will be presented ten questions per page with the same graph for reference. The chart on the left lists the *X* and *Y* coordinates you will need to find. The right chart provides answer choices A through E. Get a timer ready and give it a try.

Practice Problems

The following table is for questions 1 – 10.

X - Value

Y\X	-16	-15	-14	-13	-12	-11	-10	-9	-8	-7	-6	-5	-4	-3	-2	-1	0	1	2	3	4	5	6	7	8	9	10	11	12	13	14	15	16
16	7	3	5	2	8	10	9	9	10	11	12	12	13	14	15	15	16	17	18	19	19	20	21	22	22	23	24	25	26	26	27	28	29
15	30	29	29	28	27	26	25	25	24	23	22	21	21	20	19	18	17	17	16	15	14	13	13	12	11	10	9	9	8	9	3	5	6
14	7	23	4	5	11	7	23	39	30	29	7	23	39	29	26	15	45	7	23	9	36	30	29	7	24	3	9	36	7	23	11	21	24
13	8	14	18	47	4	8	14	20	7	23	8	14	20	31	28	7	24	8	14	23	10	7	23	55	34	6	23	10	8	14	20	55	34
12	3	9	19	46	19	3	9	15	8	14	3	9	15	32	30	55	34	3	9	29	16	8	14	9	42	2	29	16	3	9	15	9	42
11	6	23	11	98	21	6	23	28	3	9	6	23	41	34	31	9	42	6	23	14	44	3	9	6	19	1	14	44	6	23	86	6	19
10	2	29	17	58	11	16	7	18	6	23	10	17	18	13	18	6	19	8	19	66	20	6	23	11	21	29	26	15	3	9	19	5	24
9	1	14	15	54	3	14	12	13	2	29	16	16	17	18	19	20	20	21	22	23	24	2	29	26	27	31	28	7	6	23	32	32	33
8	9	11	11	44	3	9	36	5	1	14	44	48	22	43	29	44	48	22	43	7	23	1	14	11	21	32	30	55	2	29	7	23	29
7	10	9	54	24	6	23	10	11	9	11	46	50	34	44	23	46	50	34	44	8	14	9	11	38	40	34	31	9	1	14	8	14	14
6	21	5	81	4	2	29	16	17	10	9	48	51	91	44	14	48	51	91	44	3	9	10	9	22	22	22	7	24	9	11	3	9	11
5	23	13	24	45	1	14	44	15	29	26	15	45	2	3	9	44	55	17	6	4	23	13	13	12	11	15	55	34	10	9	6	23	24
4	25	33	35	36	38	40	11	11	31	28	7	24	41	6	23	23	26	28	31	33	18	7	33	23	26	28	9	42	36	38	41	55	34
3	27	19	20	21	22	22	9	54	32	30	55	34	13	2	29	16	7	18	64	22	22	12	34	16	7	18	6	19	10	17	18	9	42
2	29	14	13	13	12	11	5	81	34	31	23	11	11	1	14	7	24	44	48	22	43	13	36	14	12	13	2	29	16	16	17	6	19
1	31	28	7	24	30	29	13	24	3	22	29	17	5	9	11	55	34	46	50	34	44	15	37	13	2	29	16	16	17	20	9	54	12
0	32	30	55	34	7	23	21	23	26	28	3	9	36	38	41	9	42	48	51	91	44	17	39	23	9	81	16	15	14	13	5	81	81
-1	34	31	9	42	8	14	11	16	7	18	6	23	10	17	18	6	19	38	40	7	23	19	40	21	22	22	24	24	9	11	13	24	24
-2	7	33	25	38	3	9	3	14	12	13	2	29	16	16	17	20	21	22	22	8	14	36	22	13	12	11	34	34	10	9	7	24	24
-3	12	34	41	18	6	23	7	33	7	23	9	81	16	15	14	13	13	12	11	3	9	13	12	34	31	9	42	64	86	15	55	34	5
-4	13	36	52	39	2	29	12	34	8	14	23	24	28	31	33	3	9	36	26	6	23	9	36	3	33	3	9	36	63	89	9	42	42
-5	15	37	14	22	1	14	13	36	3	9	29	7	18	64	22	6	23	10	7	6	8	48	22	43	18	6	23	10	3	11	6	19	22
-6	17	39	21	40	9	11	15	37	6	23	14	22	26	30	7	2	29	16	24	6	46	50	34	44	8	2	29	16	35	36	30	29	31
-7	19	40	15	40	10	9	17	39	3	9	11	7	33	41	12	1	14	44	34	23	48	51	91	44	16	1	14	44	20	21	7	23	33
-8	36	22	4	41	7	33	19	40	26	10	9	12	34	41	13	36	33	35	36	38	40	20	3	9	36	23	26	28	3	9	36	38	41
-9	38	44	55	41	12	34	36	22	7	18	64	13	36	23	15	37	19	20	21	22	22	13	6	23	10	16	7	18	6	23	10	17	18
-10	40	36	7	42	13	36	5	81	3	9	36	15	37	29	17	39	14	13	13	12	11	3	2	29	16	14	12	13	2	29	16	16	17
-11	42	47	18	42	15	37	13	24	3	23	10	17	39	14	19	40	3	3	9	36	54	6	1	14	44	3	9	36	23	26	28	29	5
-12	44	48	22	43	17	39	22	1	2	29	16	19	40	11	36	22	24	6	23	10	81	2	21	23	26	28	31	33	7	24	1	14	48
-13	46	50	34	44	19	40	23	26	1	14	44	36	22	41	54	55	34	2	29	16	24	3	11	16	7	18	64	22	55	34	9	11	25
-14	48	51	91	44	36	22	16	7	18	6	23	10	17	18	81	9	42	1	14	44	28	3	9	36	38	41	43	46	9	42	10	9	1
-15	50	53	15	16	22	3	14	12	13	2	29	16	16	17	24	6	19	11	16	7	18	6	23	10	17	18	18	18	6	19	16	6	22
-16	51	27	5	45	54	15	3	54	10	11	24	20	4	89	6	55	4	2	14	12	13	2	29	16	16	17	18	19	5	20	20	4	33

Y - Value (row labels, left axis)

	X	*Y*	A	B	C	D	E
1.	−16	3	7	17	27	86	5
2.	−4	15	21	81	97	77	70
3.	14	10	23	81	73	46	19
4.	−15	1	74	40	81	62	28
5.	15	−12	39	14	25	44	13
6.	6	−5	77	12	8	22	18
7.	−10	4	90	11	13	41	58
8.	6	−9	6	82	61	82	12
9.	−5	−16	75	5	20	71	45
10.	−14	−6	66	7	94	21	55

The following table is for questions 11 – 20.

X - Value

Y\X	-16	-15	-14	-13	-12	-11	-10	-9	-8	-7	-6	-5	-4	-3	-2	-1	0	1	2	3	4	5	6	7	8	9	10	11	12	13	14	15	16
16	7	3	5	2	8	10	9	9	10	11	12	12	13	14	15	15	16	17	18	19	19	20	21	22	22	23	24	25	26	26	27	28	29
15	30	29	29	28	27	26	25	25	24	23	22	21	21	20	19	18	17	17	16	15	14	13	13	12	11	10	9	9	8	9	3	5	6
14	7	23	4	5	11	7	23	39	30	29	7	23	39	29	26	15	45	7	23	9	36	30	29	7	24	3	9	36	7	23	11	21	24
13	8	14	18	47	4	8	14	20	7	23	8	14	20	31	28	7	24	8	14	23	10	7	23	55	34	6	23	10	8	14	20	55	34
12	3	9	19	46	19	3	9	15	8	14	3	9	15	32	30	55	34	3	9	29	16	8	14	9	42	2	29	16	3	9	15	9	42
11	6	23	11	98	21	6	23	28	3	9	6	23	41	34	31	9	42	6	23	14	44	3	9	6	19	1	14	44	6	23	86	6	19
10	2	29	17	58	11	16	7	18	6	23	10	17	18	13	18	6	19	8	19	66	20	6	23	11	21	29	26	15	3	9	19	5	24
9	1	14	15	54	3	14	12	13	2	29	16	16	17	18	19	20	20	21	22	23	24	2	29	26	27	31	28	7	6	23	32	32	33
8	9	11	11	44	3	9	36	5	1	14	44	48	22	43	29	44	48	22	43	7	23	1	14	11	21	32	30	55	2	29	7	23	29
7	10	9	54	24	6	23	10	11	9	11	46	50	34	44	23	46	50	34	44	8	14	9	11	38	40	34	31	9	1	14	8	14	14
6	21	5	81	4	2	29	16	17	10	9	48	51	91	44	14	48	51	91	44	3	9	10	9	22	22	22	7	24	9	11	3	9	11
5	23	13	24	45	1	14	44	15	29	26	15	45	2	3	9	44	55	17	6	4	23	13	13	12	11	15	55	34	10	9	6	23	24
4	25	33	35	36	38	40	11	11	31	28	7	24	41	6	23	23	26	28	31	33	18	7	33	23	26	28	9	42	36	38	41	55	34
3	27	19	20	21	22	22	9	54	32	30	55	34	13	2	29	16	7	18	64	22	22	12	34	16	7	18	6	19	10	17	18	9	42
2	29	14	13	13	12	11	5	81	34	31	23	11	11	1	14	7	24	44	48	22	43	13	36	14	12	13	2	29	16	16	17	6	19
1	31	28	7	24	30	29	13	24	3	22	29	17	5	9	11	55	34	46	50	34	44	15	37	13	2	29	16	16	17	20	9	54	12
0	32	30	55	34	7	23	21	23	26	28	3	9	36	38	41	9	42	48	51	91	44	17	39	23	9	81	16	15	14	13	5	81	81
-1	34	31	9	42	8	14	11	16	7	18	6	23	10	17	18	6	19	38	40	7	23	19	40	21	22	22	24	24	9	11	13	24	24
-2	7	33	25	38	3	9	3	14	12	13	2	29	16	16	17	20	21	22	22	8	14	36	22	13	12	11	34	34	10	9	7	24	24
-3	12	34	41	18	6	23	7	33	7	23	9	81	16	15	14	13	13	12	11	3	9	13	12	34	31	9	42	64	86	15	55	34	5
-4	13	36	52	39	2	29	12	34	8	14	23	24	28	31	33	3	9	36	26	6	23	9	36	3	33	3	9	36	63	89	9	42	42
-5	15	37	14	22	1	14	13	36	3	9	29	7	18	64	22	6	23	10	7	6	8	48	22	43	18	6	23	10	3	11	6	19	22
-6	17	39	21	40	9	11	15	37	6	23	14	22	26	30	7	2	29	16	24	6	46	50	34	44	8	2	29	16	35	36	30	29	31
-7	19	40	15	40	10	9	17	39	3	9	11	7	33	41	12	1	14	44	34	23	48	51	91	44	16	1	14	44	20	21	7	23	33
-8	36	22	4	41	7	33	19	40	26	10	9	12	34	41	13	36	33	35	36	38	40	20	3	9	36	23	26	28	3	9	36	38	41
-9	38	44	55	41	12	34	36	22	7	18	64	13	36	23	15	37	19	20	21	22	22	13	6	23	10	16	7	18	6	23	10	17	18
-10	40	36	7	42	13	36	5	81	3	9	36	15	37	29	17	39	14	13	13	12	11	3	2	29	16	14	12	13	2	29	16	16	17
-11	42	47	18	42	15	37	13	24	3	23	10	17	39	14	19	40	3	3	9	36	54	6	1	14	44	3	9	36	23	26	28	29	5
-12	44	48	22	43	17	39	22	1	2	29	16	19	40	11	36	22	24	6	23	10	81	2	21	23	26	28	31	33	7	24	1	14	48
-13	46	50	34	44	19	40	23	26	1	14	44	36	22	41	54	55	34	2	29	16	24	3	11	16	7	18	64	22	55	34	9	11	25
-14	48	51	91	44	36	22	16	7	18	6	23	10	17	18	81	9	42	1	14	44	28	3	9	36	38	41	43	46	9	42	10	9	1
-15	50	53	15	16	22	3	14	12	13	2	29	16	16	17	24	6	19	11	16	7	18	6	23	10	17	18	18	18	6	19	16	6	22
-16	51	27	5	45	54	15	3	54	10	11	24	20	4	89	6	55	4	2	14	12	13	2	29	16	16	17	18	19	5	20	20	4	33

	X	Y		A	B	C	D	E
11.	6	16		22	19	32	62	21
12.	−4	11		68	87	41	81	3
13.	−11	5		14	89	68	73	50
14.	10	−3		75	42	82	44	47
15.	6	−12		53	15	25	21	85
16.	8	7		96	74	87	40	4
17.	−9	0		73	23	43	93	6
18.	−12	−3		6	14	32	30	64
19.	12	−2		45	24	18	8	10
20.	4	−5		36	8	95	40	34

The following table is for questions 21 – 30.

X - Value

	-16	-15	-14	-13	-12	-11	-10	-9	-8	-7	-6	-5	-4	-3	-2	-1	0	1	2	3	4	5	6	7	8	9	10	11	12	13	14	15	16
16	7	3	5	2	8	10	9	9	10	11	12	12	13	14	15	15	16	17	18	19	19	20	21	22	22	23	24	25	26	26	27	28	29
15	30	29	29	28	27	26	25	25	24	23	22	21	21	20	19	18	17	17	16	15	14	13	13	12	11	10	9	9	8	9	3	5	6
14	7	23	4	5	11	7	23	39	30	29	7	23	39	29	26	15	45	7	23	9	36	30	29	7	24	3	9	36	7	23	11	21	24
13	8	14	18	47	4	8	14	20	7	23	8	14	20	31	28	7	24	8	14	23	10	7	23	55	34	6	23	10	8	14	20	55	34
12	3	9	19	46	19	3	9	15	8	14	3	9	15	32	30	55	34	3	9	29	16	8	14	9	42	2	29	16	3	9	15	9	42
11	6	23	11	98	21	6	23	28	3	9	6	23	41	34	31	9	42	6	23	14	44	3	9	6	19	1	14	44	6	23	86	6	19
10	2	29	17	58	11	16	7	18	6	23	10	17	18	13	18	6	19	8	19	66	20	6	23	11	21	29	26	15	3	9	19	5	24
9	1	14	15	54	3	14	12	13	2	29	16	16	17	18	19	20	20	21	22	23	24	2	29	26	27	31	28	7	6	23	32	32	33
8	9	11	11	44	3	9	36	5	1	14	44	48	22	43	29	44	48	22	43	7	23	1	14	11	21	32	30	55	2	29	7	23	29
7	10	9	54	24	6	23	10	11	9	11	46	50	34	44	23	46	50	34	44	8	14	9	11	38	40	34	31	9	1	14	8	14	14
6	21	5	81	4	2	29	16	17	10	9	48	51	91	44	14	48	51	91	44	3	9	10	9	22	22	22	7	24	9	11	3	9	11
5	23	13	24	45	1	14	44	15	29	26	15	45	2	3	9	44	55	17	6	4	23	13	13	12	11	15	55	34	10	9	6	23	24
4	25	33	35	36	38	40	11	11	31	28	7	24	41	6	23	23	26	28	31	33	18	7	33	23	26	28	9	42	36	38	41	55	34
3	27	19	20	21	22	22	9	54	32	30	55	34	13	2	29	16	7	18	64	22	22	12	34	16	7	18	6	19	10	17	18	9	42
2	29	14	13	13	12	11	5	81	34	31	23	11	11	1	14	7	24	44	48	22	43	13	36	14	12	13	2	29	16	16	17	6	19
1	31	28	7	24	30	29	13	24	3	22	29	17	5	9	11	55	34	46	50	34	44	15	37	13	2	29	16	16	17	20	9	54	12
0	32	30	55	34	7	23	21	23	26	28	3	9	36	38	41	9	42	48	51	91	44	17	39	23	9	81	16	15	14	13	5	81	81
-1	34	31	9	42	8	14	11	16	7	18	6	23	10	17	18	6	19	38	40	7	23	19	40	21	22	22	24	24	9	11	13	24	24
-2	7	33	25	38	3	9	3	14	12	13	2	29	16	16	17	20	21	22	22	8	14	36	22	13	12	11	34	34	10	9	7	24	24
-3	12	34	41	18	6	23	7	33	7	23	9	81	16	15	14	13	13	12	11	3	9	13	12	34	31	9	42	64	86	15	55	34	5
-4	13	36	52	39	2	29	12	34	8	14	23	24	28	31	33	3	9	36	26	6	23	9	36	3	33	3	9	36	63	89	9	42	42
-5	15	37	14	22	1	14	13	36	3	9	29	7	18	64	22	6	23	10	7	6	8	48	22	43	18	6	23	10	3	11	6	19	22
-6	17	39	21	40	9	11	15	37	6	23	14	22	26	30	7	2	29	16	24	6	46	50	34	44	8	2	29	16	35	36	30	29	31
-7	19	40	15	40	10	9	17	39	3	9	11	7	33	41	12	1	14	44	34	23	48	51	91	44	16	1	14	44	20	21	7	23	33
-8	36	22	4	41	7	33	19	40	26	10	9	12	34	41	13	36	33	35	36	38	40	20	3	9	36	23	26	28	3	9	36	38	41
-9	38	44	55	41	12	34	36	22	7	18	64	13	36	23	15	37	19	20	21	22	22	13	6	23	10	16	7	18	6	23	10	17	18
-10	40	36	7	42	13	36	5	81	3	9	36	15	37	29	17	39	14	13	13	12	11	3	2	29	16	14	12	13	2	29	16	16	17
-11	42	47	18	42	15	37	13	24	3	23	10	17	39	14	19	40	3	3	9	36	54	6	1	14	44	3	9	36	23	26	28	29	5
-12	44	48	22	43	17	39	22	1	2	29	16	19	40	11	36	22	24	6	23	10	81	2	21	23	26	28	31	33	7	24	1	14	48
-13	46	50	34	44	19	40	23	26	1	14	44	36	22	41	54	55	34	2	29	16	24	3	11	16	7	18	64	22	55	34	9	11	25
-14	48	51	91	44	36	22	16	7	18	6	23	10	17	18	81	9	42	1	14	44	28	3	9	36	38	41	43	46	9	42	10	9	1
-15	50	53	15	16	22	3	14	12	13	2	29	16	16	17	24	6	19	11	16	7	18	6	23	10	17	18	18	18	6	19	16	6	22
-16	51	27	5	45	54	15	3	54	10	11	24	20	4	89	6	55	4	2	14	12	13	2	29	16	16	17	18	19	5	20	20	4	33

Y - Value (left axis)

	X	Y		A	B	C	D	E
21.	15	1		76	95	54	16	39
22.	−13	4		36	87	69	89	15
23.	12	15		23	72	62	8	40
24.	−14	−10		7	15	74	24	26
25.	−5	−13		30	36	76	27	48
26.	−2	−16		59	2	6	41	66
27.	−4	−4		29	50	45	28	59
28.	−12	9		60	25	85	59	3
29.	5	1		2	15	14	83	79
30.	3	9		38	43	23	61	34

The following table is for questions 31 – 40.

X - Value

Y	-16	-15	-14	-13	-12	-11	-10	-9	-8	-7	-6	-5	-4	-3	-2	-1	0	1	2	3	4	5	6	7	8	9	10	11	12	13	14	15	16
16	7	3	5	2	8	10	9	9	10	11	12	12	13	14	15	15	16	17	18	19	19	20	21	22	22	23	24	25	26	26	27	28	29
15	30	29	29	28	27	26	25	25	24	23	22	21	21	20	19	18	17	17	16	15	14	13	13	12	11	10	9	9	8	9	3	5	6
14	7	23	4	5	11	7	23	39	30	29	7	23	39	29	26	15	45	7	23	9	36	30	29	7	24	3	9	36	7	23	11	21	24
13	8	14	18	47	4	8	14	20	7	23	8	14	20	31	28	7	24	8	14	23	10	7	23	55	34	6	23	10	8	14	20	55	34
12	3	9	19	46	19	3	9	15	8	14	3	9	15	32	30	55	34	3	9	29	16	8	14	9	42	2	29	16	3	9	15	9	42
11	6	23	11	98	21	6	23	28	3	9	6	23	41	34	31	9	42	6	23	14	44	3	9	6	19	1	14	44	6	23	86	6	19
10	2	29	17	58	11	16	7	18	6	23	10	17	18	13	18	6	19	8	19	66	20	6	23	11	21	29	26	15	3	9	19	5	24
9	1	14	15	54	3	14	12	13	2	29	16	16	17	18	19	20	20	21	22	23	24	2	29	26	27	31	28	7	6	23	32	32	33
8	9	11	11	44	3	9	36	5	1	14	44	48	22	43	29	44	48	22	43	7	23	1	14	11	21	32	30	55	2	29	7	23	29
7	10	9	54	24	6	23	10	11	9	11	46	50	34	44	23	46	50	34	44	8	14	9	11	38	40	34	31	9	1	14	8	14	14
6	21	5	81	4	2	29	16	17	10	9	48	51	91	44	14	48	51	91	44	3	9	10	9	22	22	22	7	24	9	11	3	9	11
5	23	13	24	45	1	14	44	15	29	26	15	45	2	3	9	44	55	17	6	4	23	13	13	12	11	15	55	34	10	9	6	23	24
4	25	33	35	36	38	40	11	11	31	28	7	24	41	6	23	23	26	28	31	33	18	7	33	23	26	28	9	42	36	38	41	55	34
3	27	19	20	21	22	22	9	54	32	30	55	34	13	2	29	16	7	18	64	22	22	12	34	16	7	18	6	19	10	17	18	9	42
2	29	14	13	13	12	11	5	81	34	31	23	11	11	1	14	7	24	44	48	22	43	13	36	14	12	13	2	29	16	16	17	6	19
1	31	28	7	24	30	29	13	24	3	22	29	17	5	9	11	55	34	46	50	34	44	15	37	13	2	29	16	16	17	20	9	54	12
0	32	30	55	34	7	23	21	23	26	28	3	9	36	38	41	9	42	48	51	91	44	17	39	23	9	81	16	15	14	13	5	81	81
-1	34	31	9	42	8	14	11	16	7	18	6	23	10	17	18	6	19	38	40	7	23	19	40	21	22	22	24	24	9	11	13	24	24
-2	7	33	25	38	3	9	3	14	12	13	2	29	16	16	17	20	21	22	22	8	14	36	22	13	12	11	34	34	10	9	7	24	24
-3	12	34	41	18	6	23	7	33	7	23	9	81	16	15	14	13	13	12	11	3	9	13	12	34	31	9	42	64	86	15	55	34	5
-4	13	36	52	39	2	29	12	34	8	14	23	24	28	31	33	3	9	36	26	6	23	9	36	3	33	3	9	36	63	89	9	42	42
-5	15	37	14	22	1	14	13	36	3	9	29	7	18	64	22	6	23	10	7	6	8	48	22	43	18	6	23	10	3	11	6	19	22
-6	17	39	21	40	9	11	15	37	6	23	14	22	26	30	7	2	29	16	24	6	46	50	34	44	8	2	29	16	35	36	30	29	31
-7	19	40	15	40	10	9	17	39	3	9	11	7	33	41	12	1	14	44	34	23	48	51	91	44	16	1	14	44	20	21	7	23	33
-8	36	22	4	41	7	33	19	40	26	10	9	12	34	41	13	36	33	35	36	38	40	20	3	9	36	23	26	28	3	9	36	38	41
-9	38	44	55	41	12	34	36	22	7	18	64	13	36	23	15	37	19	20	21	22	22	13	6	23	10	16	7	18	6	23	10	17	18
-10	40	36	7	42	13	36	5	81	3	9	36	15	37	29	17	39	14	13	13	12	11	3	2	29	16	14	12	13	2	29	16	16	17
-11	42	47	18	42	15	37	13	24	3	23	10	17	39	14	19	40	3	3	9	36	54	6	1	14	44	3	9	36	23	26	28	29	5
-12	44	48	22	43	17	39	22	1	2	29	16	19	40	11	36	22	24	6	23	10	81	2	21	23	26	28	31	33	7	24	1	14	48
-13	46	50	34	44	19	40	23	26	1	14	44	36	22	41	54	55	34	2	29	16	24	3	11	16	7	18	64	22	55	34	9	11	25
-14	48	51	91	44	36	22	16	7	18	6	23	10	17	18	81	9	42	1	14	44	28	3	9	36	38	41	43	46	9	42	10	9	1
-15	50	53	15	16	22	3	14	12	13	2	29	16	16	17	24	6	19	11	16	7	18	6	23	10	17	18	18	18	6	19	16	6	22
-16	51	27	5	45	54	15	3	54	10	11	24	20	4	89	6	55	4	2	14	12	13	2	29	16	16	17	18	19	5	20	20	4	33

	X	Y	A	B	C	D	E
31.	15	−14	9	93	33	45	83
32.	−4	−9	51	1	36	41	43
33.	1	6	91	63	14	26	57
34.	−6	−9	71	37	91	50	64
35.	5	7	98	9	15	15	88
36.	−7	−5	67	11	25	9	28
37.	4	8	57	9	23	15	39
38.	14	10	19	85	92	12	43
39.	5	−16	86	2	27	76	22
40.	−11	16	45	12	88	29	10

Table Reading
Answer Key

1.	C	21.	C
2.	A	22.	A
3.	E	23.	D
4.	E	24.	A
5.	B	25.	B
6.	D	26.	C
7.	B	27.	D
8.	A	28.	E
9.	C	29.	B
10.	D	30.	C
11.	E	31.	A
12.	C	32.	C
13.	A	33.	A
14.	B	34.	E
15.	D	35.	B
16.	D	36.	D
17.	B	37.	C
18.	A	38.	A
19.	E	39.	B
20.	B	40.	E

eight

AVIATION INFORMATION

While the Aviation Information (AI) subtest consists of twenty items (questions) and lasts just eight of the 210 minutes of AFOQT testing, the amount of knowledge required to obtain a good score is substantial. The AI subtest measures the applicant's knowledge of general aeronautical concepts and terminology.

There are excellent online and printed study resources, including the Federal Aviation Administration's 281-page "Airplane Flying Handbook" (in pdf format at faa.gov). Instructional videos about aeronautics, aircraft structures and instruments, airports, and other AI topics are on YouTube.

The following information and sample multiple-choice questions will help the applicant prepare for the AI portion of the AFOQT. However, it is suggested to supplement what you find here with other expert sources if you are serious about maximizing your score. Since the AI portion of the AFOQT is so critical, we have provided double the amount of practice questions—forty in total—to help you prepare effectively.

Types of Aviation and Aircraft Categories

Since the early twentieth century, two main types of aviation have developed: **CIVIL** and **MILITARY**, both of which involve **FIXED-WING** and **ROTARY-WING** aircraft. Fighter jets, bombers, airliners, and corporate jets are examples of the former while the latter group includes helicopters, gyrocopters, and tilt-rotor flying machines.

An aircraft is a machine supported aloft by lift created by air flowing across **AIRFOIL** surfaces. The **WINGS** attached to an airplane's fuselage, a **PROPELLER** rotated by gears connected to an engine drive shaft, or spinning helicopter **ROTOR BLADES**, or by **BUOYANCY**, as in the case of airships and hot air balloons.

In the United States, civilian aircraft are certified under the following categories: normal, utility, acrobatic, commuter, transport, manned free balloons, and special classes. Special Airworthiness Certificates are issued by the Federal Aviation Administration for the following categories: primary, restricted, multiple, limited, light-sport, experimental, special flight permit, and provisional.

Military aircraft are categorized by the mission they perform: air superiority, anti-submarine warfare, coastal and sea lane patrolling, electronic warfare, ground attack (close air support), interdiction, mid-air refueling, mine sweeping, reconnaissance, search and rescue, strategic bombing, surveillance, training, transport, and weather observation.

Aircraft Structure and Components

Fixed-wing aircraft—called airplanes, or informally, planes—have a fuselage, wings, and an empennage (tail). A nose section, including the cockpit and cabin comprise the **FUSELAGE**. The **EMPENNAGE** consists of a vertical stabilizer and an attached (hinged) **RUDDER** that can be moved left or right by the pilot via cockpit controls (pedals), and a horizontal **STABILIZER** and hinged **ELEVATOR** that is also under the pilot's control and moves up and down. Some aircraft, like the U.S. Air Force's F-16 fighter jet, have an all-moving horizontal stabilizer-elevator called a **STABILATOR**.

The inboard portion of airplane wings have extendible sections called **FLAPS**, which are located along the trailing (aft) edge. They are used to increase the wing's surface area and deflect the airflow downward, thereby augmenting lift at reduced speeds. With flaps extended, planes can take off and land at a lower velocity, which requires less runway.

Some airplanes have leading-edge **SLATS**, which are also extended to maintain lift at relatively low airspeeds. Like flaps, slats help an airplane takeoff and land at a lower velocity, allowing for operations on shorter runways. The Air Force's C-17 strategic airlifter is an example of a military plane with slats and flaps.

On top of the wings of many turbine-powered aircraft are **SPOILERS**, hinged panels that move upward after landing and destroy the residual lift in order to put the plane's full weight on the landing gear and maximize tire friction on the runway, thereby enhancing deceleration.

High-performance airplanes often have one or more AIR BRAKES—also called SPEED BRAKES—to help decelerate the aircraft, and in flight, increase the rate of descent. For example, the USAF's F-15 fighter jet has a large airbrake on the top fuselage that extends after landing. Air brakes are not spoilers because they are not designed to destroy lift.

The main structural member inside each wing of an airplane is the SPAR, which runs the length of the wing. Larger wings usually have more than one spar to provide extra support. Shaped RIBS are attached perpendicularly to the spar or spars in order to provide the wing with more structure and greater strength. A SKIN of aircraft aluminum (in most cases) is attached to the framework of spar(s) and ribs.

Airplanes that fly substantially below the speed of sound typically have wings that are PERPENDICULAR to the aircraft's longitudinal (nose-to-tail) axis. The wings of most jet planes are SWEPT back to delay the drag associated with air compressibility at high subsonic speeds. Swept wings increase the performance of high-performance airplanes.

Some military aircraft have a DELTA WING (shaped like a triangle) while others have VARIABLE-GEOMETRY WINGS. In the case of the latter, the pilot swings the wings forward to a position that is roughly perpendicular to the fuselage for takeoff. This is also done when landing, flight at low airspeeds, and back when flying at high subsonic, transonic, and supersonic velocities. The Air Force's B-1B Lancer bomber is a variable-geometry airplane.

Toward the outer trailing edge of each wing is a hinged flight control surface called an AILERON that moves up and down. Ailerons operate in a direction opposite to each other and control the plane's rolling motion around the longitudinal axis. Ailerons are used to perform banking turns.

A TRIM TAB on the rear of the rudder, elevator, and one aileron (usually) act to change the aerodynamic load on the surface and reduce the need for constant pilot pressure on the control column (or joystick) and left and right pedal. Each trim tab is controlled by the pilot via a switch or wheel in the cockpit.

Regarding a source of thrust, most aircraft are powered by one or more PISTON or TURBINE ENGINES. In terms of propulsion type, the latter group consists of TURBOPROP, TURBOJET, and TURBOFAN. Fighter aircraft have one turbojet engine or a pair of them, each equipped with an AFTERBURNER, which provides an increase in thrust above non-afterburner full throttle (called military power).

The pilot controls engine operation (start, ground idle, checks, throttle movement, reverse thrust, shutdown) via switches and levers in the cockpit. The number of engine controls corresponds to the number of engines. In single- and multi-engine planes with adjustable pitch propellers, the blade angle is also controlled from the cockpit via levers.

REVERSE THRUST is a feature of turboprop and many jet-powered aircraft, including airliners, aerial tankers, and transport planes. Reverse thrust is used after landing to shorten the ground roll, the runway distance required by the decelerating airplane. Turboprop reverse thrust involves the rotation of propeller blades (three to six, typically) to a blade angle that causes air to be forced forward (away from the plane), not backward over the wings and tail surfaces, as happens when the aircraft taxis and during takeoff, climb, cruise, descent, and landing.

Reverse thrust on jet aircraft is achieved by temporarily directing the engine exhaust forward. After landing, the pilot moves the reverse thrust levers on the cockpit throttle quadrant, which causes two rounded metallic sections on the back end of each engine—called buckets, or clamshell doors—to pneumatically move and come together. When deployed, they stop the engine exhaust from going aft and direct the hot airflow forward at an angle.

Another type of reverse thrust on some jet aircraft involves pivoting doors located roughly half way along the engine. After landing, the pilot moves the reverse thrust levers, which causes the doors (four on each engine) to open. As with the buckets/clamshell doors, the result is exhaust deflected forward, which increases aircraft deceleration greatly.

Most aircraft land on wheels—called LANDING GEAR—and many types of planes have retractable wheels. Wheel retraction results in less drag when the aircraft is airborne. Fixed- and rotary-wing aircraft equipped with SKIS are able to land and maneuver on surfaces covered with snow and/or ice.

Airplanes that takeoff and land on water have FLOATS attached to supports that are connected to the fuselage, or a BOAT-LIKE HULL on the bottom of the fuselage. AMPHIBIOUS aircraft can take off and land on both land and water due to retractable wheels.

Rotary-wing aircraft (the U.S. Air Force has two fleets of them) have a FUSELAGE, TAIL, FIN (in most cases), and LANDING GEAR (e.g., skids, wheels, inflatable floats). The most common type of rotary-wing aircraft is the helicopter.

Aerodynamic Forces

There are four main aerodynamic forces that act on an aircraft when it is airborne: weight, lift, thrust, and drag.

The aircraft and everything in it—pilots, passengers, fuel, cargo, etc.—have mass (weight). Because of the earth's gravitational pull, the combined mass of the aircraft and its contents acts downward. From a physics perspective, the total weight force is deemed to act through the aircraft's CENTER OF GRAVITY.

Aerodynamic loads associated with flight maneuvers and air turbulence affect the aircraft's weight. Whenever an aircraft flies a curved flight path at a certain altitude, the load factor (force of gravity, or *G*) exerted on the airfoils (e.g., wings, rotor blades) is greater than the aircraft's total weight.

When a pilot turns an aircraft by banking (rolling) left or right, the amount of *G* increases. Banking further in order to turn more tightly causes the machine's effective weight (*G* loading) to increase more. An airplane banked 30 degrees weighs an additional 16 percent, but at 60 degrees of bank—a very steep turn—it weighs twice as much as it does in straight and level flight in smooth air.

Gusts produced by turbulent air can quickly impose aerodynamic forces that also increase the aircraft's *G* (weight) force.

LIFT is the force that counteracts an aircraft's weight and causes the machine to rise into the air and stay aloft. Lift is produced by airfoils that move through the air at a speed sufficient to create a pressure differential between the two surfaces and a resulting upward force. Lift acts perpendicular to the direction of flight through the airfoil's **CENTER OF PRESSURE** or **CENTER OF LIFT**.

THRUST is an aircraft's forward force, which is created by one or more engines (the largest plane in the world, the Antonov An-225 Mriya, has six huge turbofan jet engines). In propeller-driven airplanes and rotary-wing aircraft, the power output of the engine(s) is transformed into rotary motion via one or more transmissions (gear boxes). Generally, thrust acts parallel to the aircraft's longitudinal axis.

DRAG opposes thrust; it is a rearward-acting force caused by airflow passing over the aircraft's structure and becoming disrupted. Drag acts parallel to the **RELATIVE WIND** and is a function of aircraft shape and size, its velocity and angle (inclination) in relation to airflow, and the air's mass, viscosity, and compressibility.

An aircraft's **TOTAL DRAG** is the sum of its **PROFILE DRAG**, **INDUCED DRAG**, and **PARASITE DRAG**. When total drag is the lowest, the aircraft experiences its maximum endurance (in straight and level flight), best rate of climb, and for helicopters, minimum rate-of-descent speed for autorotation.

PROFILE DRAG is the sum of **FORM DRAG** and **SKIN FRICTION**. Form drag varies with air pressure around the aircraft and its cross-sectional shape. Skin friction is a function of the roughness of the outer surface of an aircraft (due to surface imperfections, protruding rivet heads, etc.).

INDUCED DRAG is a product of lift; stationary aircraft generate no such drag. However, as lift is created during acceleration along the runway or strip (in the case of airplanes) or increased rotor RPM and angle of attack (in the case of helicopters), the resulting pressure differential between the airfoil surfaces creates an air vortex at the wing's or rotor blade's tip. The vortex moves parallel to the aircraft's longitudinal axis and expands in diameter with distance from the airfoil. The effect of each vortex is a retarding aerodynamic force called induced drag.

Parts of an aircraft that do not contribute to the production of lift create **PARASITE DRAG** when the machine is moving. On airplanes, such components include the nose section and fuselage, landing gear, engine pylons and cowlings, vertical stabilizer, and rudder. On helicopters, the cockpit and cabin, landing skids or wheels, externally mounted engines (on some types), tail boom, and fin create parasite drag.

Scientific Principles of Relevance to Aeronautics

Bernoulli's Principle

In 1738, a Swiss scientist named Daniel Bernoulli published a book entitled *Hydrodynamica*. In it he explained that an increase in the inviscid flow of a fluid (i.e., the flow of an ideal, zero-viscosity liquid or gas) resulted in a decrease of static pressure exerted by the fluid. Bernoulli's famous equation is as follows:

$$P + \frac{1}{2}\rho v^2 = \text{a constant}$$

P = pressure (a force exerted divided by the area exerted upon); ρ (the Greek letter "rho") = the fluid's density; and v = the fluid's velocity.

The constant in Bernoulli's formula is derived from the scientific principle that energy cannot be created or destroyed—only its form can be changed—and a system's total energy does not increase or decrease.

Conservation of Energy

Bernoulli's Principle is based on the conservation of energy. It says that in a steady flow the sum of all forms of mechanical energy—a fluid's potential energy plus its kinetic energy—along a streamline (e.g., a tube) is the same at all points. Thus, greater fluid flow rate (higher speed) results in increased kinetic energy and dynamic pressure and reduced potential energy and static pressure.

An aircraft filled with fuel has a finite amount of energy. Through combustion in the engine, the fuel's heat energy is converted into kinetic energy. Either in the form of jet exhaust or at least one rotating propeller (many types of planes have two or more propellers). Spinning helicopter rotor blades also have kinetic energy.

If an aircraft is airborne when it runs out of fuel, it still has potential energy as a function of its height above the ground. As the pilot noses down to keep air flowing over the airfoils (wings, rotor blades) and create lift, the aircraft's potential energy is transformed into kinetic energy.

Combining Bernoulli's Principle with the fact that airfoils provide lift at varying speeds during different phases of flight (takeoff, climb, cruise, descent, landing), the lift produced at a given instant can be calculated using the following equation:

$$L = \tfrac{1}{2}\rho v^2 A C_1$$

L = the lift force, $\tfrac{1}{2}\rho v^2$ was previously explained, A = the airfoil's area (length multiplied by width), and C_1 is the coefficient of lift of the airfoil.

Pilots need to remember that the lifting force on their aircraft is proportional to the density (ρ) of air through which they fly (higher altitude = less dense air), the aircraft's speed, and airfoil angle of attack (AOA).

Conservation of Mass

In the scientific field of fluid dynamics, it has been established that a fluid's mass cannot be created or destroyed within a flow of interest (e.g., airflow in sub-zero temperature conditions). Conservation of mass is mathematically expressed as the mass continuity equation.

Conservation of Momentum

Momentum, an object's mass times its velocity, cannot be created or destroyed. However, it can be changed through an applied force. Because it involves magnitude and direction, momentum is a vector quantity. It is conserved in all three directions (longitudinally, laterally, and in terms of yaw) simultaneously.

Venturi Effect

To understand how a machine with airfoils can take to the air and remain airborne, we need to examine a phenomenon called the Venturi Effect. In the late eighteenth century, an Italian physicist, Giovanni Battista Venturi, conducted experiments with a pump and an unusual tube. The diameter of one end of the tube was constant, and the circumference of the tube's central portion was smaller. Downstream from the bottleneck, the tube's diameter increased. It was as though someone had squeezed the center of the tube, creating a constriction.

Venturi noticed that as fluids moved through the tube, the flow rate increased (accelerated) and the force (static pressure) against the tube's surface decreased as the diameter became smaller. The opposite phenomenon—reduced flow rate (deceleration) and greater static pressure—happened as the tube diameter downstream of the constriction widened. Venturi published his findings in 1797 and the effect that he observed, measured, and wrote about became associated with his name. It has certainly been integral to aviation since the advent of gliding centuries ago.

If a Venturi tube is cut in half longitudinally, the curvature of the tube wall will look similar to that of an airplane wing's upper surface or the top of helicopter rotor blades. A moving airfoil "slices" the air, forcing molecules to travel along one side or the other. Those moving across the curved side have to travel a greater distance to reach the trailing edge than those moving across the relatively flat side. Consequently, the air molecules moving across the curved surface accelerate, as they did in Venturi's tube, and the static pressure drops.

Because pressure flows from high to low, the static pressure differential experienced between an airfoil's two sides imposes an aerodynamic force acting from the high-pressure (flat) surface to the low-pressure (curved) side. When acting upward, the force is called lift.

Newton's First Law of Motion

A stationary object remains at rest and an object in motion continues to move at the same rate (speed) and in the same direction unless acted upon by a force.

Newton's Second Law of Motion

Acceleration results from a force being applied to an object. The heavier the body, the greater the amount of force needed to accelerate it.

Newton's Third Law of Motion

Sir Isaac Newton (1642 – 1727) was a brilliant English physicist and mathematician who formulated universal laws of motion, including his third, which said that for every action there is an equal and opposite reaction. Consequently, when an airfoil is deflected up, the airstream flowing over the airfoil reacts by moving downward. Also, when exhaust from jet engines is directed backward the resulting reactive force on the engines, engine pylons, wings, and the rest of the airplane is forward.

Aircraft Axes

Aircraft motion occurs around three axes—longitudinal, lateral, and yaw—that go through the machine's center of gravity. The longitudinal axis has been explained; an aircraft rolls around it. The lateral axis is horizontal and perpendicular to the longitudinal axis; on basic airplane images it is depicted as a straight line going through one wingtip to the other. The yaw axis is vertical; an aircraft is said to yaw (rotate) around it.

Flight control

When lift = weight and thrust = drag, the aircraft is either stationary on the ground, or aloft in straight-and-level, unaccelerated flight. To make an aircraft accelerate requires an increase in thrust, which the pilot controls from the cockpit by moving one or more throttle controls (on piston aircraft) or power lever(s) on turbine aircraft.

During takeoff, the airplane accelerates along the runway, strip, or body of water and reaches a speed at which it is going fast enough for the wings to generate lift. To make the plane go skyward, the pilot pulls back on the control column, or joystick, which causes (via cables in lighter, smaller aircraft, or a hydraulic system in larger, heavier planes) the hinged elevator to tilt up.

The inclined elevator forces air passing over it to deflect up, resulting in a downward reaction force on the airplane's tail. Because the elevator is aft of the aircraft's center of gravity, the tail drops as the nose of the plane rises and the aircraft climbs.

To make the plane descend, the opposite happens.

To turn an airplane, the pilot moves the control wheel or joystick to the left or right (as desired) to change the machine's direction. The aileron on the wing on the plane's side to where the pilot wants to turn rises into the airstream. This forces the flow upward and reduces the lift produced by the outer portion of the wing where the aileron is located. The result is a wing that drops, rolling (banking) the aircraft.

On the opposite side of the airplane, the aileron moves down into the airstream, deflecting the airflow downward and creating more lift, which causes the wing to rise. With one wing down and the opposite wing up, the airplane rolls to the left or right.

For a coordinated banked turn, the pilot needs to move the aircraft's rudder to the side of the turn (left, right), which is accomplished by pushing on the corresponding pedal in front of him or her. As the pilot does so, the airflow passing over the moved rudder is deflected to the left or right, corresponding to the pushed pedal. The reactive force against the vertical stabilizer is opposite (right, left) and because the tail is aft of the plane's center of gravity, the nose yaws around the yaw axis in the opposite direction (left, right).

Aviation Terms and Definitions

Airfoil: A wing or helicopter blade that generates more lift than drag as air flows over its upper and lower surfaces. A propeller is also an airfoil. Airfoils are carefully designed and can be made of non-metallic materials such as composites.

Angle of attack: The angle between the chord line of an airfoil and its direction of motion relative to the air (i.e., the relative wind). AOA is an aerodynamic angle.

Angle of incidence: In the context of fixed-wing airplanes, the angle of incidence is the inclination of the wing or tail surface attached to the fuselage relative to an imaginary line that is parallel to the aircraft's longitudinal axis.

ANHEDRAL ANGLE: The downward angle of an airplane's wings and tailplane from the horizontal is called the anhedral angle, or negative dihedral angle.

ATTITUDE: An aircraft's position relative to its three axes and a reference such as the earth's horizon.

CENTER OF GRAVITY (CG): An aircraft's center of mass, the theoretical point through which the entire weight of the machine is assumed to be concentrated.

CHORD: The distance between the leading and trailing edges along the chord line is an airfoil's chord. In the case of a tapered airfoil, as viewed from above, the chord at its tip will be different than at its root. Average chord describes the average distance.

CHORD LINE: An imaginary straight line from the airfoil's leading (front) edge to its trailing (aft) edge.

CONSTANT SPEED PROPELLER: A controllable-pitch propeller whose angle is automatically changed in flight by a governor in order to maintain a constant number of revolutions per minute (RPM) despite changing aerodynamic loads.

CONTROLLABILITY: A measure of an aircraft's response relative to flight control inputs from the pilot.

CONTROLLABLE PITCH PROPELLER: A propeller that can be varied in terms of its blade angle by the pilot via a control in the cockpit.

COORDINATED FLIGHT: When the pilot applies flight and power control inputs to prevent slipping or skidding during any aircraft maneuver, the flight is said to be coordinated.

CRITICAL ANGLE OF ATTACK: The angle of attack at which an airfoil stalls (loses lift) regardless of the aircraft's airspeed, attitude, or weight.

DIHEDRAL ANGLE: The upward angle of an airplane's wings and tailplane from the horizontal.

DIHEDRAL EFFECT: The amount of roll moment produced per degree of sideslip is called dihedral effect, which is crucial in terms of an aircraft's rolling stability about its longitudinal axis.

DIRECTIONAL STABILITY: An aircraft's initial tendency about its yaw (vertical) axis. When an aircraft is disturbed yaw-wise from its equilibrium state due to a gust, for example, and returns to that state (i.e., aligned with the relative wind) because of the aerodynamic effect of the vertical stabilizer, it is said to be directionally stable.

DOWNWASH: Air that is deflected perpendicular to an airfoil's motion.

DRAG COEFFICIENT: A dimensionless quantity that represents the drag generated by an airfoil of a particular design.

DRAG CURVE: A constructed image of the amount of aircraft drag at different airspeeds.

DYNAMIC STABILITY: Describes the tendency of an aircraft after it has been disturbed from straight-and-level flight to restore the aircraft to its original condition of flying straight and level by developing corrective forces and moments.

EQUILIBRIUM: In the context of aviation, equilibrium is an aircraft's state when all opposing forces acting on it are balanced, resulting in unaccelerated flight at a constant altitude.

FEATHERING PROPELLER: A controllable-pitch propeller that can be rotated sufficiently by the pilot (via a control lever in the cockpit connected to a governor in the propeller hub) so that the blade angle is parallel to the line of flight, thereby minimizing propeller drag.

FORWARD SLIP: A pilot-controlled maneuver where the aircraft's longitudinal axis is inclined to its flight path.

GLIDE RATIO: The ratio between altitude lost and distance traversed during non-powered flight (e.g., following an engine failure, in a sailplane).

GLIDEPATH: An aircraft path's across the ground while approaching to land.

GROSS WEIGHT: An aircraft's total weight when it is fully loaded with aircrew, fuel, oil, passengers and/or cargo (if applicable), weapons, etc.

GYROSCOPIC PRECESSION: The attribute of rotating bodies to manifest movement ninety degrees in the direction of rotation from the point where a force is applied to the spinning body.

HEADING: The direction in which the aircraft's nose is pointed.

INERTIA: A body's opposition to a change in motion.

INTERNAL COMBUSTION ENGINE: A mechanical device that produces power from expanding hot gasses created by burning a fuel-air mixture within the device.

Lateral stability (rolling): An aircraft's initial tendency relative to its longitudinal axis after being disturbed, its designed quality to return to level flight following a disturbance such as a gust that causes one of the aircraft's wings to drop.

Lift coefficient: A dimensionless quantity that represents the lift generated by an airfoil of a particular design.

Lift/drag ratio: A number that represents an airfoil's efficiency, the ratio of the lift coefficient to the drag coefficient for a specific angle of attack.

Lift-off: The act of rising from the earth as a result of airfoils lifting the aircraft above the ground.

Load factor: The ratio of the load supported by an aircraft's lift-generating airfoils (wings, main rotor blades) to the aircraft's actual weight, including the mass of its contents. Load factor is also known as *G*-loading (*G* means gravity).

Longitudinal stability: An aircraft's initial tendency relative to its lateral axis after being disturbed, its designed quality to return to its trimmed angle of attack after being disrupted due to a wind gust or other factor.

Maneuverability: An aircraft's ability to change directions in three axes along its flight path and withstand the associated aerodynamic forces.

Mean camber line: An imaginary line between the leading and trailing edges and halfway between the airfoil's upper (curved) and lower (flat) surfaces.

Minimum drag speed (L/DMAX): The point on the total drag curve where total drag is minimized and lift is maximized (i.e., where the lift-to-drag ratio is greatest).

Nacelle: An enclosure made of metal or another durable material that covers an aircraft engine.

Non-symmetrical airfoil (cambered): When one surface of an airfoil has a specific curvature that the opposite side does not, the airfoil is described as non-symmetrical, or cambered. The advantage of a non-symmetrical wing, for example, is that it produces lift at an AOA of zero degrees (as long as airflow is moving past the blade). Moreover, the lift-to-drag ratio and stall characteristics of a cambered airfoil are better than those of a symmetrical airfoil. Its disadvantages are center of the pressure movement chord-wise by as much as one-fifth the chord line distance, which causes undesirable airfoil torsion, and greater production costs.

NORMAL CATEGORY: An airplane intended for non-acrobatic operation that seats a maximum of nine passengers and has a certificated takeoff weight of 12,500 pounds or fewer.

PAYLOAD: In the context of aviation, the weight of an aircraft's occupants, cargo, and baggage.

P-FACTOR (PRECESSION FACTOR): A propeller-driven aircraft's tendency to yaw to the left when the propeller rotates clockwise (as seen by the pilot) because the descending propeller blade on the right produces more thrust than the ascending blade on the left. If the propeller rotated counter-clockwise, the yaw tendency would be to the right.

PISTON ENGINE: Also known as a reciprocating engine, it is a heat engine that uses one or more pistons to convert pressure created by expanding, hot gases resulting from a combusted fuel-air mixture, or steam pressure, into a rotating motion.

PITCH: An airplane's rotation about its lateral axis, or the angle of a propeller blade as measured from the vertical plane of rotation.

POWER LEVER: The cockpit lever connected to a turbine engine's fuel control unit, which changes the amount of fuel entering the combustion chambers.

POWERPLANT: An engine and its accessories (e.g., starter-generator, tachometer drive) and the attached propeller (usually via a gearbox).

PROPELLER BLADE ANGLE: The angle between the chord of an airplane propeller blade and the propeller's plane of rotation.

PROPELLER LEVER: The cockpit control that controls propeller speed and angle.

PROPELLER SLIPSTREAM: Air accelerated behind a spinning propeller.

PROPELLER: A relatively long and narrow blade-like device that produces thrust when it rotates rapidly. In aviation, the term typically includes not only the propeller blades but also the hub and other components that make up the propeller system.

RATE OF TURN: The rate of a turn expressed in degrees per second.

RECIPROCATING ENGINE: An engine that converts heat energy created by combusted fuel mixed with air into reciprocating piston movement, which in turn is converted into a rotary motion via a crankshaft.

REDUCTION GEAR: A gear or set of gears that turns a propeller at a speed slower than that of the engine.

RELATIVE WIND: The direction of airflow relative to an airfoil, a stream of air parallel and opposite to an aircraft's flight path.

RUDDERVATOR: Two control surfaces on an aircraft's tail that form a *V*. When moved together via the control wheel or joystick in the cockpit, the surfaces act as elevators. When the pilot presses his or her foot against one rudder pedal or the other, the ruddervator acts like a conventional plane's rudder.

SIDESLIP: A flight maneuver controlled by the pilot that involves the airplane's longitudinal axis remaining parallel to the original flight path, but the aircraft no longer flies forward, as in normal flight. Instead, the horizontal lift component causes the plane to move laterally toward the low wing.

SKID: A flight condition during a turn where the airplane's tail follows a path outside of the path of the aircraft's nose.

SLIP: A maneuver used by pilots to increase an aircraft's rate of descent or reduce its airspeed, and to compensate for a crosswind during landing. An unintentional slip also occurs when a pilot does not fly the aircraft in a coordinated manner.

STABILITY: An aircraft's inherent tendency to return to its original flight path after a force such as a wind gust disrupts its equilibrium. Aeronautical engineers design most aircraft to be aerodynamically stable.

STALL: A rapid decrease in lift caused by an excessive angle of attack and airflow separating from an airfoil's upper surface. An aircraft can stall at any pitch attitude or airspeed.

STANDARD-RATE TURN: A rate of turn of three degrees per second.

SUBSONIC: Speed below the speed of sound, which varies with altitude.

SUPERSONIC: Speed in excess of the speed of sound, which varies with altitude.

SWEPT WING: A wing planform involving the tips being further back than the wing root.

SYMMETRICAL AIRFOIL: When an airfoil has identical upper and lower surfaces, it is symmetrical and produces no lift at an AOA of zero degrees. The wings of very high performance aircraft tend to be symmetrical.

TAXIWAY LIGHTS: Blue lights installed at taxiway edges.

TAXIWAY TURNOFF LIGHTS: Green lights installed level with the taxiway.

THROTTLE: A mechanical device that meters the amount of fuel-air mixture fed to the engine.

THRUST LINE: An imaginary line through the center of an airplane's propeller hub and perpendicular to the propeller's plane of rotation, or through the center of each jet engine.

TOTAL AERODYNAMIC FORCE (TAF): Two components comprise the total aerodynamic force: lift and drag. The amount of lift and drag produced by an airfoil are primarily determined by its shape and area.

TORQUE: A propeller-driven airplane's tendency to roll in the opposite direction of the propeller's rotation. Some multi-engine airplanes have propellers that rotate in opposite directions to eliminate the torque effect.

TRAILING EDGE: The aft part of an airfoil where air that was separated as it hit the wing's front edge and was forced over the upper and lower surfaces come together.

TRANSONIC: At the speed of sound, which varies with altitude.

TRIM TAB: A small, hinged control surface on a larger control surface (e.g., aileron, rudder, elevator) that can be adjusted in flight to a position that balances the aerodynamic forces. In still air, a trimmed aircraft in flight requires no control inputs from the pilot to remain straight and level.

T-TAIL: The description for an airplane's tail involving the horizontal stabilizer mounted on the top of the vertical stabilizer.

TURBULENCE: The unsteady flow of a fluid (e.g., air).

UTILITY CATEGORY: An airplane intended for limited acrobatic operation that seats a maximum of nine passengers and has a certificated takeoff weight of 12,500 pounds or fewer.

VECTOR: A force applied in a certain direction.

Depicted visually, a vector shows the force's magnitude and direction.

Velocity: The rate of movement (e.g., miles per hour, knots) in a certain direction.

Vertical stability: An aircraft's designed, inherent behavior relative to its vertical axis, its tendency to return to its former heading after being disturbed by a wind gust or other disruptive force. Also called yawing or directional stability.

V-tail: A design involving two slanted tail surfaces that aerodynamically behave similarly to a conventional elevator and rudder, i.e., as horizontal and vertical stabilizers.

Wing: An airfoil attached to a fuselage that creates a lifting force when the aircraft has reached a certain speed.

Wing area: A wing's total surface, including its control surfaces, and winglets, if so equipped.

Wing in ground effect (WIG): When an aircraft flies at a very low altitude, one roughly equal to its wingspan, it experiences WIG. The effect increases as the airplane descends closer to the surface (runway, land, water) and supports the aircraft on a cushion of air best at an altitude of one half the wing span.

Winglet: A surface installed on a wingtip that is angled to the wing and improves its efficiency by smoothing the airflow across the upper wing near the tip and reducing induced drag. Winglets improve an aircraft's lift-to-drag ratio.

Wing span: The maximum distance between wingtips.

Wingtip vortices: A spinning mass of air generated at a wing's tip created by outward-flowing high pressure air from underneath the wing meeting inward-flowing low air pressure on the wing's upper surface. The intensity of a wing vortex—also referred to as wake turbulence—is dependent on an airplane's weight, speed, and configuration.

Wing twist: A wing design feature that improves the effectiveness of aileron control at high angles of attack during an approach to a stall.

Practice Problems

1. A propeller-driven airplane—

   1-A   is part of the rotary class of aircraft (because the propeller spins)

   1-B   has a reciprocating engine only

   1-C   is a fixed-wing aircraft

   1-D   has a reverse thrust feature in all types of military and civilian aircraft

2. Military aircraft are categorized—

   2-A   as normal, utility, acrobatic, special mission, or transport

   2-B   based on the mission they perform

   2-C   in accordance with Department of Defense directives since 1947

   2-D   none of the above

3. A propeller is—

   3-A   an airfoil

   3-B   a secondary source of thrust

   3-C   part of a balanced thrust system involving only 2, 4, or 6 blades

   3-D   an extendible thrust-generation device used at high altitudes

4. The four main forces acting on an aircraft are—

   4-A   deflection, exponential thrust, torque, and the total mass vector modified by the Earth's Coriolis Effect

   4-B   lift, weight, thrust, and drag

   4-C   wind gusts, gravity, pressure differentials, and tangential rotation

   4-D   all of the above

5. Turbine aircraft—

   5-A   have a propeller source of thrust in some cases

   5-B   never have a propeller source of thrust (only jets are turbine aircraft)

   5-C   have a turbocharged engine

   5-D   utilize a ducted wind fan that spins an electrical generator

6. The empennage consists of—

   6-A   a vertical stabilizer and a hinged rudder.

   6-B   the back half of the fuselage and the tailplane.

   6-C   the *T* tail and nacelle

   6-D   the ruddervator and associated hydraulic system

7. Flaps are used—

   7-A   to decrease Dutch roll

   7-B   to eliminate wingtip vortices

   7-C   during takeoff only

   7-D   none of the above

8. Hinged wing panels that move upward and destroy lift after landing are called—

   8-A   air brakes

   8-B   dpoilers

   8-C   winglets

   8-D   vertical stabilizers

9. Swept-back wings—

   9-A   delay the drag associated with air compressibility at approach speeds

   9-B   delay the drag associated with air compressibility at low subsonic speeds

   9-C   delay the drag associated with air compressibility at high subsonic speeds

   9-D   all of the above

10. Profile drag is the sum of—

    10-A   Skin friction and form drag

    10-B   Skin friction and induced drag

    10-C   Form drag and supplementary drag

    10-D   Parasite drag and vortex drag

11. Slats are located—

   11-A   along the horizontal stabilizer's leading edge

   11-B   along the leading edge of both wings and the horizontal stabilizer

   11-C   along the trailing edge of the right aileron

   11-D   along the leading edge of both wings

12. The laws of conservation that pertain to aircraft are—

   12-A   the law of conservation of mass, kinetic energy, and fluid flow

   12-B   the law of conservation of mass, torque, and potential energy

   12-C   the law of conservation of weight, thrust, and lift

   12-D   the law of conservation of mass, energy, and momentum

13. According to 18th century Swiss scientist Daniel Bernoulli:

   13-A   Accelerated fluid flow results in a decrease in dynamic pressure.

   13-B   Accelerated fluid flow results in a decrease in static pressure.

   13-C   Accelerated fluid flow results in an increase of total system energy.

   13-D   all of the above

14. Lift produced by an airfoil is proportional to—

   14-A   the rate of air compressibility and the coefficients of lift and drag

   14-B   the angle of airflow deflection, the relative wind's vertical vector component, and the reduction of induced drag as the aircraft accelerates

   14-C   air density, aircraft speed, wing area, and airfoil shape

   14-D   none of the above

15. The angle of attack is—

   15-A   The chord line's orientation in relation to the aircraft's longitudinal axis.

   15-B   The acute angle between the chord line of an airfoil and the relative wind.

   15-C   The sum of the angle of incidence of the wings and tailplane.

   15-D   The aircraft's downward inclination when shooting targets on the ground.

16. Parasite drag is produced by—

   16-A   extended slats and flaps

   16-B   aircraft parts that do not contribute to producing lift

   16-C   improperly set trim tabs

   16-D   a difference in propeller RPM on multi-engine airplanes

17. Thrust opposes—

   17-A   drag

   17-B   rudder deflection

   17-C   gyroscopic precession

   17-D   gravity

18. Ailerons move—

   18-A   in opposing directions

   18-B   downward

   18-C   up or down, depending on the rudder pedal pushed by the pilot

   18-D   none of the above

19. The main types of turbine propulsion are—

   19-A   axial and centrifugal flow

   19-B   non-afterburning, after burning, and turbocharged

   19-C   turbofan, turbojet, and turboprop

   19-D   turbocharged, turbofan, and ramjet

20. An aircraft's three axes are—

20-A longitudinal, gyroscopic, and lateral

20-B directional, pitch, and gyroscopic

20-C yaw, longitudinal, and lateral

20-D deflectional, lateral, and induced

21. Fill in the blanks below:

Increasing an aircraft's bank in a coordinated turn, _____ its _____ and _____.

21-A increases; angle of attack; lift

21-B increases; weight (due to G loading); rate of turn

21-C decreases; angle of attack; drag.

21-D decreases; weight (due to G loading); angle of attack.

22. When the pilot pulls back on the control column or joystick:

22-A The elevator moves up.

22-B The elevator moves down.

22-C The left aileron moves down.

22-D none of the above

23. To move the rudder to the right, the pilot—

23-A turns the control wheel to the right

23-B pulls back on the right power lever

23-C moves the right throttle lever forward while pushing the right pedal

23-D pushes the right pedal

24. Fill in the blank below:

From a physics perspective, an aircraft's total weight force is deemed to act through the _____.

24-A weight and balance reference datum

24-B center of pressure

24-C center of gravity

24-D center of momentum

25. An air vortex at the wing tip creates—

25-A form drag

25-B profile drag

25-C induced drag

25-D parasite drag

26. Momentum is—

26-A an object's mass times its velocity squared

26-B an object's mass times its velocity

26-C an object's weight plus one-half of its velocity squared

26-D an object's forward velocity times its coefficient of lift

27. Fill in the blank below:

When exhaust from jet engines is directed backward, the resulting reactive force on the airplane is _____.

27-A forward

27-B forward but deflected downward due to the angle of incidence

27-C forward but reduced because of the inclined component of the total drag vector

27-D determined only by using the conservation of energy equation

28. Coordinated flight is defined as—

28-A the pilot applying control inputs that are suitable for the aircraft's density altitude

28-B the pilot applying flight and power control inputs to prevent slipping or skidding during any aircraft maneuver

28-C the pilot reducing back pressure on the control column or joystick while turning in the opposite direction of the horizontal component of total drag

28-D all of the above

29. Fill in the blank below:

Anhedral angle is the _____ angle of an airplane's wings and tailplane from the horizontal:

29-A   upward

29-B   obtuse

29-C   downward

29-D   isoceles

30. Minimum drag speed corresponds to—

30-A   the point on the total drag curve where the thrust-to-drag ratio is least

30-B   the point on the total drag curve where the drag-to-mass ratio is least

30-C   the point on the total drag curve where the lift-to-drag ratio is greatest

30-D   the point on the total drag curve where the lift-to-weight ratio is least

31. An airfoil stalls when:

31-A   The downward component of the wingtip vortices are greater than the lift produced by increasing the angle of attack.

31-B   There is a rapid decrease in lift caused by an excessive angle of attack and airflow separating from an airfoil's upper surface.

31-C   The pilot has mistakenly extended the flaps while flying above the maneuvering airspeed (Va).

31-D   The pilot deploys the air brakes.

32. Fill in the blank below:

A propeller with a blade angle that can be changed by the pilot is called a _____ propeller.

32-A   dynamic

32-B   rotational

32-C   reverse thrust

32-D   controllable

33. The attribute of rotating bodies to manifest movement ninety degrees in the direction of rotation from the point where a force is applied to the spinning body is called—

33-A   rotational precession

33-B   dynamic precession

33-C   induced precession

33-D   gyroscopic precession

34. An aircraft's initial tendency relative to its longitudinal axis after being disturbed and dropping a wing to return to level flight is known as—

34-A   lateral stability

34-B   longitudinal stability

34-C   directional stability

34-D   none of the above

35. An imaginary line from an airfoil's leading edge to its trailing edge that is halfway between the airfoil's upper and lower surfaces is the—

35-A   mean camber line

35-B   chord line

35-C   angle of incidence

35-D   elevator inclination line

36. When one surface of an airfoil has a specific curvature that the opposite side does not have, the airfoil is described as—

36-A   non-cambered

36-B   deflected

36-C   non-symmetrical

36-D   laterally torqued

37. The phenomenon of a propeller-driven aircraft's tendency to yaw to the left when the propeller rotates clockwise (as seen by the pilot) because the descending propeller blade on the right produces more thrust than the ascending blade on the left is known as—

37-A   asymmetric thrust

37-B   rotational precession

37-C   p-factor (precession factor)

37-D   directional instability

38. Airflow parallel and opposite to an aircraft's flight path is called the—

   38-A   relative wind

   38-B   longitudinal wind

   38-C   dynamic wind

   38-D   none of the above

39. The speed of sound varies with—

   39-A   angle of attack

   39-B   angle of inclination

   39-C   induced drag

   39-D   altitude

40. A propeller-driven airplane tends to roll in the opposite direction of the propeller's rotation because of—

   40-A   the induced plane of rotation

   40-B   tangential drag

   40-C   torque

   40-D   angular momentum

Aviation Information
Answer Key

1.	C	21.	B
2.	B	22.	A
3.	A	23.	D
4.	B	24.	C
5.	A	25.	C
6.	A	26.	B
7.	D	27.	A
8.	B	28.	B
9.	C	29.	C
10.	A	30.	C
11.	D	31.	B
12.	D	32.	D
13.	B	33.	D
14.	C	34.	A
15.	B	35.	A
16.	B	36.	C
17.	A	37.	C
18.	A	38.	A
19.	C	39.	D
20.	C	40.	C

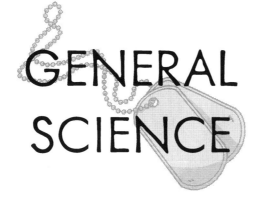

nine

GENERAL SCIENCE

The purpose of this section of the AFOQT is to determine how well the applicant understands various (basic) concepts from the physical sciences, the life sciences, and Earth sciences. This is a section which is basically going to be all rote memorization on your part. That can be a daunting task, but it shouldn't be something which gives you pause. You will do just fine if you go through this material a couple of times. Most of it, in fact, will probably seem like common sense if you remember your basic science classes from the past.

One thing to keep in mind is that there are a lot of facts and figures which will show up in this section. You will see a ton of questions that cover everything from physics to chemistry. Rather than trying to memorize all of the little individual facts that you come across, it will pay for you to try and memorize the base foundations. This would be all of the general principles that are behind the little facts and figures. Consider it akin to looking at the "big picture" of the scientific world. That will send you on the way to a good score.

On the General Sciences subtest, you have eleven minutes for twenty questions. That is not a lot of time for each individual question, as you can probably gather immediately upon looking at those figures. The bright side, though, is that there is not a lot of interpretation that goes on in this section. Either you know the correct answer to a question or you do not know the correct answer to a question. It is as simple as that. Most of the questions that you will encounter here are going to take less than ten seconds to answer, and you have about thirty-three seconds on average.

Study Information

One of the first things you need to know, and something that will apply to every single subsection you will find below, is how measurement works in the scientific world. The system, which differs significantly from the system used in the United States, is known as the metric system.

Table 9.1. Metric system units of measure

Physical Property	Basic Unit	Symbol
time	second	s
volume	liter	L
mass	gram	g
length	meter	m

These units can be modified using prefixes based on the power of ten. For example, 1000 meters is equal to one kilometer.

Table 9.2 Metric system prefixes

Prefix	Abbreviation	Multiples
tera	T	10^{12}
giga	G	10^9
mega	M	10^6
kilo	k	10^3
hecto	h	10^2
deka	da	10
deci	d	10^{-1}
centi	c	10^{-2}
milli	m	10^{-3}
micro	u	10^{-6}
nano	n	10^{-9}
pico	p	10^{-12}
femto	f	10^{-16}
atto	a	10^{-18}

If you are wondering how these two tables fit together, think about this example.

What is a ks? We know from the first table that *k* means kilo and that kilo means 1000 (10^3). From the second table, we know that *s* means second, the basic metric unit of time. So *ks* is a kilosecond or, 1000 seconds.

So what about temperature? Most people know there are at least two temperature scales which are commonly used. The Celsius [C] scale and the Fahrenheit [F] scale. There is also a third, the Kelvin scale. This scale comes into play during high-level chemistry and physics discussions. Now the three scales will be broken down, and conversions between C and F will be discussed.

- **KELVIN:** This scale is a bit more specific than the other two. It is used to help determine what the absolute coldest temperature possible is. On this scale, absolute zero is defined as the temperature at which molecular motion would cease to occur. That temperature is 0 K. For a bit of reference, the freezing point of water would be 273.15 K.

- **FAHRENHEIT:** The standard temperature scale in the United States. Water freezes at 32°F and water boils at 212°F.

- **CELSIUS:** The metric standard scale for temperature across the world. The freezing point of water is 0°C, and the boiling point is 100°C.

There are two types of conversion systems. One is a bit more complex than the other.

Table 9.3. Conversion systems

	SYSTEM ONE	SYSTEM TWO
From °F to °C	$C = \frac{5}{2} \times (F - 32)$	$C = [\frac{5}{2} \times (40 + F)] - 40$
From °C to °F	$F = \frac{2}{5} \times (C + 32)$	$F = [\frac{2}{5} \times (40 + C)] - 40$

This may seem complicated, but you have to bear in mind that the two resulting equations are the exact opposite of each other insofar as the multiplication is concerned, so if you know one of them you can easily figure out the other.

The Scientific Method

One of the things that sets the modern sciences apart from the methods done in the past is a system of hypothesis testing known as the scientific method. Through this, new ideas can be cultivated and tested in order to help form the basis for the various sciences.

Here are the basic steps of the scientific method:

1. **OBSERVE** things going on around you or in the universe in general.
2. **FORM A HYPOTHESIS** (an educated guess) about why your observation occurs.
3. **PREDICT AN OUTCOME** based on that hypothesis.
4. **EXPERIMENT** using the hypothesis and prediction as a basis and record your results. If they do not match, modify the hypothesis.
5. **REPEAT STEPS 3 AND 4** until you are able to come up with a repeatable result.

Disciplines

A number of scientific disciplines are covered on the AFOQT.

Table 9.4. AFOQT scientific disciplines

SPECIALTY	DESCRIPTION
agriculture	the study of farming
archeology	the study of past anthropology. Civilizations, tools, etc.
astronomy	study of space
biology	the study of life
botany	the study of plant life
chemistry	the study of chemicals, elements, and reactions
ecology	the study of the environment
entomology	the study of insects
genealogy	the study of ancestry
genetics	the study of genes and heredity
geology	the study of rocks and minerals
ichthyology	the study of fish, a sub-specialty of biology
meteorology	the study of weather
paleontology	study of prehistoric life

All of these individual disciplines are covered by the subheadings below. Knowing the niche specialties can be important, however, so don't skimp here. With that being said, the questions won't be as simple as "what does a meteorologist do".

Life Sciences

Life sciences cover exactly what you might think that it covers. Biology and systems related to it. This includes everything from the smallest microbe to the largest organism and everything in between. It also includes the classification systems that we use to help piece together the bigger picture of biology.

Classifications

Classifications are how scientists denote differences and relationships between different organisms. Every single organism is going to have its own designation (usually two words, the first is the GENUS and the second is the SPECIES). Most of these names are derived from Latin, so a basic familiarity with that language will help a lot here (and with nearly all of the other sciences as well).

The following is a list of the classification system (all eight levels):

1. domain   5. order
2. kingdom   6. family
3. phylum   7. genus
4. class   8. species

To see a full example of this in action, let's look at humans:

Table 9.5. Human classification

domain	eukaryota
kingdom	animalia
phylum	chordata
class	mammalia
order	primates
family	hominidae
genus	homo
species	homo sapien

As you can see, the classification gets more and more specific the closer to the species it gets. Finally, you arrive at the smallest distinction that you can get to at the species level.

As far as kingdoms are concerned, most scientists agree that there are around 5:

- FUNGI are multi-celled organisms that do not undergo photosynthesis. Mushrooms belong to this kingdom.

- PROTISTS are single-celled organisms which have a nucleus.

- MONERANS AND VIRUSES are single-celled, non-nucleated, organisms. This includes bacteria, algae, and viruses. Viruses are sometimes argued to be their own kingdom, as they are unique in the world of biology.

- PLANTS is one of the two largest kingdoms (animals is the other). This includes non-moving organisms with no obvious sensory or nervous systems. They have cell walls consisting of cellulose.

- ANIMALS are multi-celled organisms without cellulose-comprised cell walls, chlorophyll, or the ability to undergo photosynthesis. Organisms in this kingdom can both respond to external stimuli and can move.

Evolution

EVOLUTION is a term which describes a change in characteristics of a population of organisms over a given period of time. The process is usually slow and gradual. One important thing to note is that evolution does not have an end goal. It is a process which is constantly ongoing and constantly changing. There is no beginning or end to the process.

These principles can be used to help sum up the way evolution works:

- A huge amount of genetic variation exists among different organisms of the world.
- Organisms have to compete for a limited supply of resources.
- The organisms which can survive the best and are able to reproduce will be naturally selected for continuation because of that.

Two important things to note with regard to natural selection: genetic variation is random among individuals, and traits allowing an organism to survive and reproduce are going to be passed down to the offspring of that organism. Traits that do the opposite will die off. Organisms which are adapted better to their particular environment are going to be more likely to survive and, thus, their traits will be more likely to live on in subsequent generations.

Some factors which can influence and contribute to the process of evolution include the following:

- **GENETIC DRIFT**: When groups leave larger populations and establish new populations, they can become genetically different from their parent populations, leading to speciation (at times).
- **MIGRATION**: This occurs when individuals move into and out of a given population, allowing gene flow between multiple populations.
- **MUTATION**: The process of genetic material replicating itself is not perfect. In fact, there are systems built into the process which cause mutations in order to potentially adapt organisms better to their environments. These can be harmful or beneficial to the individual.

So how do scientists know that evolution is real? How do they know this is something which actually happens? Ever since the publication of *On The Origin of Species* by Charles Darwin, scientists have been collecting evidence and striving to prove the theory. One of the largest places from which they derive their evidence is through the fossil record, which shows the way modern organisms have descended from common ancestors (think birds from dinosaurs).

Another common way this is done is through comparative anatomy. Many different types of organisms have similar structures. The bones of the arms of a human, the wings of a bird, or the fins of a porpoise, for example, all share some common bones and traits, though the exact shape may differ amongst them.

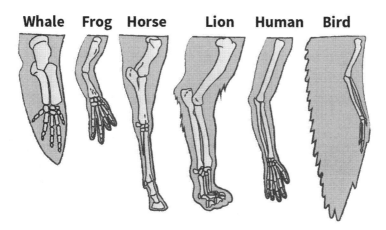

Whale   Frog   Horse   Lion   Human   Bird

Figure 9.1. Homologous structures

The above image shows some common structures across different organisms.

Embryology is another area which is commonly studied when looking into evolution. The embryos of many animals look nearly the same until they begin to differentiate, suggesting that they all come from the same ancestor organism. Chicken embryos and human embryos are a particularly striking example of this.

Finally, the study of genetics and biochemistry has greatly enhanced the ability of scientists to look at evolutionary and genetic changes on a very small scale (at the DNA and RNA level).

The Cell

Cells are the most basic unit of every living organism. Some organisms are, in fact, just single cells. Others, like the complex organisms that humans, animals, and most plants are, are comprised of multiple cells which are both specialized and organized into groups that make up tissues and organs. Every single cell is going to have two basic things: a PLASMA MEMBRANE and CYTOPLASM. The plasma membrane is the outermost boundary of the cell. This is what keeps the inside of the cell inside and keeps the outside of the cell outside. Cytoplasm is the liquid-like substance which makes the basic foundation of the cell. This is located inside of the plasma membrane and helps to give the cell a shape.

Cells are either PROKARYOTIC or EUKARYOTIC. Prokaryotic cells are usually very simple, like bacterial cells. These are almost universally going to be single celled organisms and will not comprise any substantial parts of multicellular organisms. Eukaryotic cells are more complex and contain specialized organelles within them which carry out special functions.

The primary components of eukaryotic cells are NUCLEI, MITOCHONDRIA, RIBOSOMES, and CHLOROPLASTS:

- The NUCLEUS is where genetic information is stored and accessed within the cell.

- The MITOCHONDRIA is where ATP is produced, the primary energy molecule in most cells. This is colloquially known as the "powerhouse" of the cell.
- RIBOSOMES are where cells make proteins.
- CHLOROPLASTS are found only in plant cells; this is the site of photosynthesis. This is where plant cells make their energy and convert it into sugar for storage.

It is worth noting that there are many other organelles in most eukaryotic cells. These, however, are the most basic and most important ones. These are the ones you are most likely to encounter on the AFOQT.

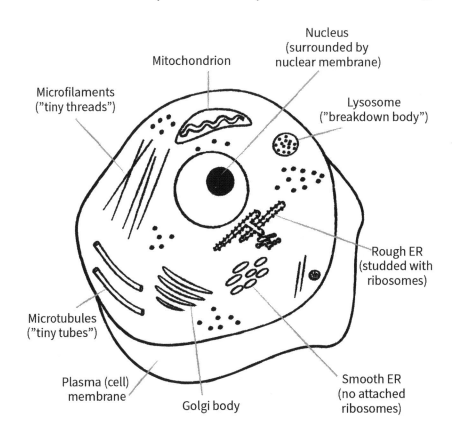

Figure 9.2. Eukaryotic cell

The plasma membrane, which you will remember is the outermost boundary of the cell, is studded with proteins and other mechanisms which help with the selective permeability which is required for substances to move in and out of the cell. There are four primary ways through which things can pass inside and outside of plasma (cell) membranes:

1. ACTIVE TRANSPORT: Molecules are able to cross the plasma membrane going from regions of low concentration to regions of high concentration using specialized proteins located within the membrane itself. This process requires the expenditure of energy reserves (ATP).

2. FACILITATED DIFFUSION: Specialized proteins allow diffusion across the cell membrane during the right conditions.

3. OSMOSIS: This is a specialized form of diffusion which concerns itself with the movement of water across the cell membrane, usually to equalize pressure or change relative concentration.

4. DIFFUSION: This is the passive movement of molecules from regions of high concentration to regions of low concentration (think about putting a drop of food coloring into a glass of water).

Respiration and Fermentation

The way that many organisms receive energy is known as the process of cellular respiration. During this process, cells break down carbohydrate molecules, usually glucose through a process known as glycolysis (*glyco* means sugar, *lysis* means to break apart). The energy produced through cellular respiration is stored as adenosine triphosphate (ATP). When cells require energy, they break down the ATP and then utilize the energy from the breaking of the bond. Respiration requires the presence of oxygen. It takes place within the mitochondria.

The process of cellular respiration can be summarized as:

$$\text{glucose (sugar)} + O_2 \text{ (oxygen)} > \text{water} + CO_2 \text{ (carbon dioxide)} + \text{ATP (energy)}$$

When oxygen is not available, cells of a few organisms will switch to a process known as anaerobic (*an* means without; *aerobic* means oxygen) respiration. This process is also known as fermentation. The products of this process are ethanol and carbon dioxide.

Photosynthesis

Photosynthesis is the process through which plants make the food that they use for energy. They use sunlight, carbon dioxide, and water in the process of photosynthesis to make glucose. The process requires energy, which is where the sunlight comes into play. Sunlight is captured in specialized organelles (chloroplasts) which use pigments (chlorophyll) to capture different wavelengths of light.

The process of photosynthesis can be summarized as:

$$CO_2 \text{ (carbon dioxide)} + \text{water} > \text{glucose (sugar)} + O_2 \text{ (oxygen)} + \text{water}$$

Interestingly enough, plants are green because they do not use green light and, instead, they reflect it back.

Cell Division

The cells of living organisms have the ability to reproduce exact copies of themselves through division. This, along with the expansion of cells, is how organisms grow and develop.

MEIOSIS is the process through which an organism creates four daughter cells from the parent cell. Each of these will have half of the number of chromosomes that the parent cell has. These daughter cells are gametes (eggs and sperm) which are used for sexual reproduction.

MITOSIS is the process most cells go through. Two copies of the original cell are created. There are a number of phases to this process. Each of the two daughter cells will have the same number of chromosomes as the parent cell has.

Genetics

Genetics is an enormous subject. It is growing larger every single day, and the body of knowledge is absolutely staggering. With that being said, only the basics of genetics are going to be covered on the AFOQT.

Genetics is the study of genes and how they relate to the traits and function of organisms. Genes themselves are small parts of DNA molecules which relate to specific functions and traits in the larger individual to which they belong. Each gene can have more than one form. These individual forms are known as alleles. Which allele an organism inherits from parent organisms is how traits manifest themselves.

Here are some common terms which you should know for the AFOQT:

- **TRANSCRIPTION** is the process through which DNA is copied onto RNA.
- **TRANSLATION** is the process through which RNA is translated into proteins.
- **GENOTYPE** is the sum total of the genes in an individual.
- **PHENOTYPE** is the way those genes express themselves in characteristics of an individual.
- The **RECESSIVE ALLELE** is the allele which is not expressed when they are paired with a dominant allele. Geneticists usually use a lower case letter to represent the alleles which are recessive.
- If two different alleles are present, and one of them is expressed, then that is the **DOMINANT ALLELE** of the two. This is usually going to be shown as a capital letter.
- **HETEROZYGOTE** is an individual who has two different alleles for the same gene.
- **HOMOZYGOTE** is an individual who has two identical copies of the same gene.

The CENTRAL DOGMA of molecular biology is that the flow of information flows from DNA > RNA > protein.

Note: These definitions are very simplistic. They will suit you for the AFOQT, but you may want to brush up on them individually in order to get a better understanding of the core concepts.

Human Anatomy and Reproduction

The study of anatomy and physiology should start with diet and nutrition. This is something which is often overlooked in many textbooks. In addition, it is a somewhat confusing topic because of the "buzzword" nature of the whole thing. Everyone knows about fad diets and the latest food trends. Not many people, however, know much about the basics of what foods are composed of.

Here is a brief rundown of nutrients obtained from food sources:

- VITAMINS are rganic compounds which are essential to health. Generally, these are going to be proteins. Some break down easily in fats, and some break down easily in water.

- MINERALS are elements required by animals to build new proteins and to function properly. Think phosphorous, magnesium, calcium, and zinc (among many others, of course).

- LIPIDS form cell membranes, and they help to comprise some hormones. Fat is used as a storage source for energy as well.

- NUCLEIC ACIDS are used to form RNA and DNA, nucleic acids are gained from eating other nucleic acids (and then subsequently breaking them down).

- CARBOHYDRATES are the basic source of energy all animals use. Glucose is the most commonly used carbohydrate for energy, and it is created during the process of cellular respiration.

- PROTEINS form the framework and most of the structures found in the body. They are comprised of amino acids which are gained by eating other protein sources.

The Digestive System

The human digestive system is composed of a number of parts which work together in conjunction. An important aspect of the digestive system is the assistance of a variety of gut bacteria in the process of digestion.

Here are some of the major components of the human digestive system and their function:

- The **MOUTH** is a familiar organ to most people. This is where you break large food particles into smaller ones through a process known as mastication (chewing). Saliva also helps to begin the digestion process.
- The **ESOPHAGUS** is a highly muscular tube which carries food from the mouth down to the stomach. Food is moved through a rhythmic contraction of the muscles located there (the process known as peristalsis).
- The **STOMACH** is an expandable bag-like organ inside the abdomen. It is filled with gastric juices (and acids) which break down food. The food is then moved to the small intestine.
- The **SMALL INTESTINE** is where absorption of food occurs, primarily.
- The **LARGE INTESTINE** is where remaining waste is eliminated and turned into feces which can then be passed out through the anus to leave the body.

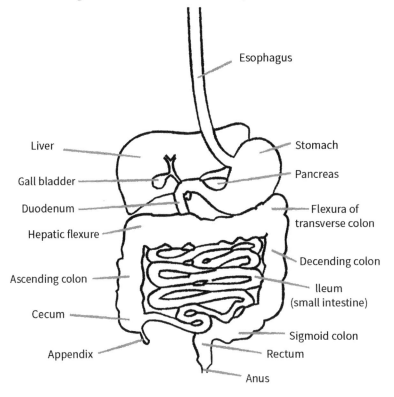

Figure 9.3. Digestive system

The Respiratory System

The respiratory system consists primarily of the following organs:

- The **LUNGS** consist of small sacs called alveoli; this is where capillaries from the circulatory system undergo gas exchange, releasing carbon dioxide, and binding with oxygen.

- The TRACHEA is where air branches into the two lungs.
- The PHARYNX passes air from the nose into the throat.
- The NOSE is where the respiratory system begins. Air is moistened and otherwise conditioned here. Hairs in the nose help to trap dust and to purify the air.

The Excretory System

This is the system that gets rid of the waste products that the body makes. The human kidneys are two organs located near the stomach and the liver. This is where blood is filtered. Once it has been filtered, waste products are carried through ureters (tubes) to the urinary bladder where it is stored until it can be released. Urine is composed of uric acid and urea, among other things, which are the waste products resulting from the breakdown of nucleic acids and amino acids, respectively.

The Nervous System

The human nervous system is comprised of two sub-systems. The central nervous system and the peripheral nervous system.

The CENTRAL NERVOUS SYSTEM consists of the brain and the spinal cord. Of the two, this is the most important. The central nervous system is where sensory information is processed and where your consciousness, intelligence, and memories are stored.

Extending from the base of the brain down to the base of the spine, the SPINAL CORD connects the peripheral nervous system to the brain. Impulses and signals travel up and down the spinal cord, which acts as a center of coordination for the nervous system as a whole.

- The cerebrum, thalamus, hypothalamus, and limbic system make up the FOREBRAIN. The forebrain controls sensory input, learning, memory, emotion, and the synthesis of hormones.
- Between the fore- and hindbrains, the MIDBRAIN consists primarily of nerves and fibers used to alert the forebrain should something out of the ordinary occur.
- The medulla, cerebellum, and pons are what make up the HINDBRAIN. The hindbrain helps to bridge the regions of the brain and to help coordinate muscle movements.

The PERIPHERAL NERVOUS SYSTEM is all of the other nerves in the human body. These are the nerves that connect the central nervous system to all of your peripheral parts.

- The AUTONOMIC NERVOUS SYSTEM is comprised of the sympathetic and the parasympathetic nervous systems. This is an involuntary series of nerves which help to prepare the body for emergency situations (fight or flight, for example) and helps with sleeping and preparation for digestion.

- The sensory somatic system is the system which carries any impulses from the environment and the senses, which allows people to be both aware of their environment and to act on the information they are receiving.

The Circulatory System

The human circulatory system is the system which transports oxygen and other nutrients to the tissues of the body from the lungs and the intestines.

Here are some of the most important components of the circulatory system:

- **BLOOD** is comprised of plasma, platelets, red blood cells, and white blood cells.
- Red blood cells are also known as **ERYTHROCYTES**. These are donut shaped cells with no nucleus. They carry **HEMOGLOBIN**, which gives them their red color, and binds oxygen and carbon dioxide to carry it to and from body tissues.
- White blood cells are known as **LEUKOCYTES**. These are a major part of the human immune system.
- **PLATELETS** are small and disc-shaped fragments of cells which are created in the bone marrow. These are the starting material that is used to initiate the blood clotting process.
- **PLASMA** is a straw yellow colored fluid like substance comprised of other blood proteins and constituents, including waste products, hormones, and nutrients. Its primary constituent is water.
- **LYMPH** is a watery fluid which is derived from blood plasma and is an active component of the immune system.
- The **HEART** is the organ which keeps your blood flowing through your body. It does this through a rhythmic series of contractions which occur involuntarily.

Ecology

Ecology is a specialty of biology which concerns itself with interactions between the environment and organisms in that environment. Here are some basic definitions you will need to know in order to navigate this section of the AFOQT:

- An **ECOSYSTEM** includes every species and every organism which is living in a certain community and how they interact with each other. This includes things such as soil, water, and light as well.
- A **POPULATION** is members of a single species which are living in a pre-determined area of the environment.

- A **COMMUNITY** is the specific living area of animal and plant species.

There are many ways that organisms in a specific ecosystem may interact with one another. **PREDATION** is when one one organism feeds on another one. One of the organisms in the situation will benefit while the other is injured in some way. **PARASITISM** is when one organism benefits and the other is harmed. In parasitism, one organism will live off of or feed on another, usually without killing it outright. **COMMENSALISM** is when one organism benefits while the other is not affected one way or another. Lastly, **MUTUALISM** is when both organisms benefit from this type of relationship.

Here is a basic layout of the ways that energy flows through a given ecosystem, being transferred from one place to another:

- **PRODUCERS** are the organisms which trap the energy of the sun through photosynthesis (e.g., plants and algae).
- **PRIMARY CONSUMERS** are the herbivores which feed on the producers of the environment.
- **SECONDARY CONSUMERS** are usually carnivores. These are the organisms which feed on the primary consumers of the environment.
- **DECOMPOSERS** are the organisms which help to break down dead organisms and then recycle their nutrients into the environment.

The study of ecology is important, but it usually takes a backseat to more formal studies of the individual subject areas of life science. With that being said, ecology is becoming more and more prevalent as a field of study given the rise in environmental awareness all around the world.

Chemistry

Chemistry is one of the physical sciences and is one of the basic sciences that form the general science subtest of the AFOQT. For this subtest, you will be concerned with the composition and structure of matter, including the properties of matter and the way that matter changes over time and due to environmental factors. The primary concern of chemistry is with atoms and molecules, both the way that they interact with one another and the transformations that they can undergo.

Chemistry is arguably the most critical science. Nearly every other scientific discipline needs to utilize information from chemistry in order to move forward with their specific research. It also helps that chemistry is the bridge between physics, biology, and the other sciences. It is a type of physical science, but it should not be confused with physics.

Atomic Structure the Nature of Matter

So if chemistry is basically at matter, what, exactly *is* matter? MATTER is anything that has a mass and that takes up a given volume. The three most common states of matter are all things which you are familiar with already: gasses, liquids, and solids.

- SOLIDS have a defined shape, volume, and mass.
- LIQUIDS have a defined volume and a defined mass, but they do not have a definite shape.
- GASSES have a definite mass but no volume or shape. They expand to fill whatever they are put in.

The state of matter that a particular element or molecule is found in is determined by the amount of kinetic energy it has. KINETIC ENERGY is the energy of motion. All particles of matter are moving. They are constantly moving. The speed at which they are moving is what primarily determines their state of matter. Gasses move the fastest, and then liquids, then solids.

Temperature can play a role in this. When you heat up a solid, it will become a liquid because the particles speed up. When you heat a liquid, it becomes a gas. When you apply cold, the opposite happens. Heat speeds up the motion of molecules, while cold slows down the motion of molecules.

There is another form of matter which bears mentioning as well: plasma. PLASMA is an ionized state of matter which is somewhat similar to a gas. Don't expect to see many questions about plasma on the AFOQT—what questions you do see will most likely not be anything complex.

Atoms, Molecules, and Compounds

The basic building block of everything in chemistry is the atom. The atom is generally defined as the smallest unit of matter which defines a chemical element. Every type of matter is made of up one or more atoms, either neutral in charge or ionized (with a charge). These are very small particles, measured in picometers.

Atoms are composed of a few subatomic (smaller than atoms) building blocks. The first of these is the nucleus, which is the core of the atom. Electronics are negatively charged particles which circle around (outside of) the nucleus. Inside of the nucleus are protons (positive charge) and neutrons (neutral charge). The vast majority of the mass of an atom is within the nucleus. It should also be noted that most atoms are going to be of neutral charge. There will be as many electrons as there are protons, so they will cancel out each other's charge. Additionally, there are usually as many neutrons as there are protons. If the number of neutrons differs from the number of protons, then that atom is known as an ISOTOPE. You may be familiar with this concept through terms

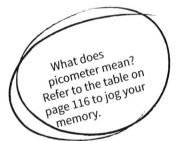

What does picometer mean? Refer to the table on page 116 to jog your memory.

like CARBON **13** or something similar. Isotopes are frequently used to track molecules and atoms moving through systems. Below is a table outlining the most common subatomic particles and some of their characteristics.

Table 9.6. Subatomic particles

PARTICLE	CHARGE	MASS IN GRAMS	SYMBOL
electron	−1	9.109×10^{-28}	e
neutron	0	1.675×10^{-24}	n
proton	+1	1.673×10^{-24}	p

For example, hydrogen, the simplest atom, contants 1(+1) proton, 0 neutrons, and 1(-1) electrons.

The electrons that are located outside of the nucleus form a "cloud" or a number of SHELLS (the most common way of describing them, and essential to advanced forms of chemistry).

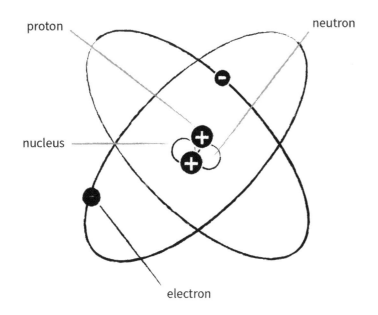

Figure 9.4. Structure of an atom

Given all of the different combinations of protons and electrons, there is an almost infinite number of atoms which could be formed. At the current time, however, only 122 are known. There are others that have been theorized, and some which have been created artificially, but only 122 are known to exist.

The next thing that you need to concern yourself with is molecules. You must know by now that everything is not made up of just individual atoms. Different combinations of atoms go together to form molecules and compounds. Compounds are substances which have more than one type of atom bound together with it. Molecules are compounds which are formed of tightly bound nonmetals (this will make more sense after the discussion of the periodic table). That is a distinction which can be somewhat confusing, but it important to remember.

There are a number of elements which are always found in nature as molecules. These are known as **DIATOMIC MOLECULES**. Some of these elements are only found in nature either in compounds with other elements or bound to another element of the same type. Here are the most common diatomic molecules in nature:

- oxygen (O_2)
- fluorine (F_2)
- hydrogen (H_2)
- nitrogen (N_2)

Compounds have properties which differ from the individual elements from which they are composed. This is evident, readily, in the case of, say, hydrogen (H_2) versus water (H_2O). At room temperature, hydrogen is going to be a gas. At room temperature, water is going to be a liquid. Obviously not the same, but both of them nevertheless have hydrogen as a constituent part.

The Periodic Table

The periodic table has been designed in order to easily display elements, along with their most common characteristics, in one easy to read table. In this table, elements are assigned either a one-letter or a two-letter designation which is known as their atomic symbol. For example hydrogen is represented by the letter H, helium is represented by the letters He, and so on. The first letter is always going to be capitalized and the second letter is always going to be lower case. This is a method of keeping elements straight when they are written out with other elements in chemical formulas (more on this later).

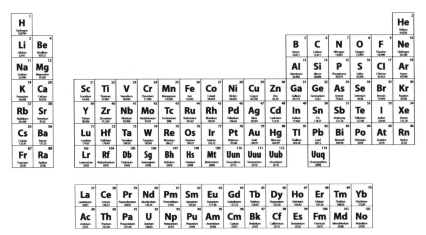

Figure 9.5. Periodic table of elements

At first glance, this table may appear to be confusing or nonsensical. Once you understand the method by which it was designed, however, you will quickly discover that it is able to hold an enormous amount of information in a relatively small space.

The periodic table is read from left to right, and the atoms on the table have been arranged by the number of protons they have within their nucleus. This is also known as the ATOMIC NUMBER. Hydrogen has one proton, so it comes first on the table. It should be noted that at a certain point, the number of protons between atoms on the table is no longer a difference of one. For example, the difference in Ra and Lr is 88 and 103, and they are found right beside each other on the table. The atomic number (the number of protons) is the characteristic that defines an atom of a particular element. All atoms of the same element have to have the same number of protons. They do not, however, have to have the same number of neutrons or electrons.

If an atom has the same amount of protons and electrons, then the atom is going to be neutral. The positive and negative charges of that atom will cancel each other out. If there are fewer electrons than there are protons, then the atom will be a positively charged CATION. In the opposite case, an atom with more electrons than protons is going to be a negatively charged ANION. Isotopes are neutrally charged atoms with a different number of protons than neutrons, which has already been discussed.

The ATOMIC WEIGHT (atomic mass) of an element is going to be listed on the periodic table right beneath the symbol. The atomic mass is going to be the average mass of all of the isotopes of an element which can naturally occur. Additionally, the columns of the table indicate the group number for individual elements. The groups indicate how many electrons are in the outermost shell of an atom. These outermost electrons are commonly called VALENCE ELECTRONS. They are the electrons which hold different atoms together when they form molecules.

The table itself was given its name because of the trends in the elements that appear when they are arranged in order of atomic number. This shows differences between elements which are regular (periodic) but are not readily apparent. This includes delineations between nonmetals and metals in addition to other things (semimetals, halogens, etc.).

Reactions and Equations

Studying elements and molecules is great, but how is it going to help you understand the chemical processes which are going on all the time? The answer to that question is in the study of reactions and chemical equations. Chemical reactions and equations are the shorthand which chemists use to describe chemical processes.

On the following page is an example of a basic chemical equation which can be used to illustrate the basic concepts:

$$C + O_2 > CO_2$$

This reaction is a **COMBUSTION REACTION** which sees carbon and oxygen combine to form carbon dioxide. The two reactants, oxygen and carbon, of the equation are found on the left-hand side of the arrow. The arrow itself is what represents the reaction occurring. The right-hand side of the equation shows the product(s) of the reaction. In this case, that product is carbon dioxide. The **LAW OF CONSERVATION OF MASS** states that matter can be neither created nor destroyed, only modified. In this situation, it means that you need to have the same elements and the same number of atoms on both sides of the reaction. As you can see in this example, there is one carbon atom on both sides and two oxygen atoms on both sides. The equation has been balanced.

There are a few types of chemical reactions that you can expect to encounter on this subtest of the AFOQT. A **DECOMPOSITION REACTION** is a reaction in which a substance breaks down into multiple constituent components (at least two). Below is the formula for this type of reaction, as well as an example of a common decomposition reaction:

$$AB > A + B$$

$$2H_2O > 2H_2 + O_2$$

A **SYNTHESIS REACTION** (combination reaction) is a reaction in which at least two substances react with each other to form a single compound. Below is the formula for this type of reaction, as well as an example of a common synthesis reaction:

$$A + B > AB$$

$$3H_2 + N_2 > 2NH_3$$

A **SINGLE REPLACEMENT/DISPLACEMENT REACTION** is a reaction in which a single element reacts with a compound and exchanges places with one of the component elements of that compound. Below is the formula for this type of reaction, as well as an example of a common single replacement/displacement reaction:

$$AB + C > AC + B$$

$$3AgNO_3 + Al > Al(NO_3)_3 + 3Ag$$

A **DOUBLE REPLACEMENT/DISPLACEMENT REACTION** is a reaction in which two compounds react in order to achieve an exchange. Below is the formula for this type of reaction, as well as an example of a common double replacement/displacement reaction:

$$AB + CD > AD + CB$$

$$AgNO_3 + NaCl > AgCl + NaNO_3$$

In the above reactions, notice that the same elements in the same quantities are found on the product side and the reaction side. Mass is neither created nor destroyed. All of the equations have been balanced.

Both types of replacement/displacement reactions are collectively known as substitution reactions.

Solutions, Bases, Acids

Solutions, bases, and acids are some of the most common types of chemicals you will encounter in your day to day life. The following are some definitions to help get you started:

- The **pH SCALE** measures how much acid is in a solution (the power of hydrogen in the solution). A pH between 0 and 7 is acidic. A pH of exactly 7 is neutral. A pH between 7 and 14 is basic.

- An **ACID** is any compound that raises the amount of hydrogen (H+) ions in a solution.

- A **BASE** is any compound that decreases the H+ concentration in a solution by increasing the concentration of hydroxide (OH–).

- A **NEUTRAL SOLUTION** is a solution which is neither acidic nor basic. pH of 7 is neutral.

- A **SOLUTION** is a homogeneous mixture which is composed of two things: a solvent and a solute.

- A **SOLVENT** is the material which is in larger proportion within the solution (the one doing the dissolving).

- A **SOLUTE** is the material which is being dissolved in the solvent.

- A **BUFFER** is a substance which helps to ease the change in pH when acids or bases are added to a solution.

If you are a bit confused as to the definitions between a solution, solvent, and solute, take the example of salt water. Salt water is a solution in which the water acts as the solvent and the salt (sodium chloride [NaCl]) acts as the solute.

Water, in fact, is known as the universal solvent because of how many different substances it is capable of dissolving. Water is neutral (when pure). Below are some of the most common bases you have likely encountered:

- **SODIUM CARBONATE**, Na_2CO_3, is used in the manufacturing of paper and is also used to "soften" water.

- **LIME**, CaO, is typically used to raise the pH of the soil for growing certain kinds of crops. It is a commonly used substance in the farming industry.

- **AMMONIA**, NH_3, is a chemical which is used commonly in household cleaners. It is also used in fertilizers. It is one of the primary constituents of urine.

- **LYE** (sodium hydroxide), NaOH, is a commonly used base which is used in order to make soap.

- ACETONE, C_3H_6O, is commonly used as a solvent. It is often used for stripping paint.

Below are some common examples of acids:

- SULFURIC ACID, H_2SO_4, is the chemical compound with the highest amount of industrial production in the entire world. It is also utilized in the creation and operation of car batteries.

- PHOSPHORIC ACID, H_3PO_4, is commonly used in soft drinks in order to help stem potential bacterial growth that could occur in them. It is also used in the production of many commercial fertilizers.

- VINEGAR (acetic acid), HC_2H3O_2, is a solution of acetic acid (usually 5 percent acetic acid in water).

- CITRIC ACID, $H_3C_6H_5O_7$, is found in fruits. It is what gives them their slightly acidic and tangy taste. In addition, it is utilized in the metabolism of many animals

- CARBONIC ACID, H_2CO_3. This acid is the acid which everyone talks about being in carbonated soft drinks. The acid is formed when carbon dioxide is dissolved in water.

- NITRIC ACID, HNO_3, is a type of acid that is commonly used in the creation of fertilizers (due to the nitrogen content).

- HYDROCHLORIC ACID, HCl, is a very common acid. You may also know it as stomach acid.

The Most Important Elements

Of all of the elements, the first twenty which are listed in the periodic table are the most important. They are the most abundant on Earth, and they are likely the most abundant in the entire universe. The twenty elements listed below are used in materials that people use every single day, in both industries and in their own bodies. The format for these elements is as follows: name (symbol/atomic number): information.

1. HYDROGEN (H/1): Hydrogen is the most abundant element in the universe. It is clear, odorless, and low density. It is also highly flammable, which is why it is no longer commonly used for blimps and balloons. It is found in nature as a diatomic element (H_2). Hydrogen ions which have been dissolved in water cause that water to become acidic.

2. HELIUM (HE/2): Helium is a clear gas with no odor and low density. It is also the second most abundant element found in the universe. It is not, however, one of the most abundant elements found on Earth. Helium is not flammable, so it is commonly utilized in lighter-than-air applications such as blimps and balloons. Helium is not

particularly reactive, and it does not occur as a diatomic molecule when found in elemental form.

3. **LITHIUM (LI/3)**: Lithium is a metal with a low density. It is reactive when it is found on its own. It forms a +1 ion very easily. One of the most common uses of elemental lithium is in the treatment of some psychological disorders (particularly bipolar disorders).

4. **BERYLLIUM (BE/4)**: Beryllium is a metal with a low density. It is commonly used in technology due to the strength that it imparts. At the same time, it is very toxic to humans, so care must be taken when it is handled.

5. **BORON (B/5)**: Boron is a gas at room temperature. It is classified as a metalloid, which has some of the properties of metals. When boron is turned into an oxide, it is used to make some cleaning agents and is a major constituent of heat resistant glassware.

6. **CARBON (C/6)**: Carbon may be the most important element that there is, right after hydrogen. Carbon is a solid when it is found at room temperature. It is also an extremely versatile element. Graphite, the substance used to make pencil lead, and diamonds are both forms of carbon. Carbon is capable of forming four bonds with other elements (or with itself). Carbon is also the basis of nearly all known life forms and is the major distinguishing element between organic and inorganic chemistry. Carbon dioxide is one of the major products of fossil fuel burning and, as such, is one of the commonly cited greenhouse gasses that environmentalists and policy makers concern themselves with.

7. **NITROGEN (N/7)**: Nitrogen, in its natural form, is an odorless and clear gas with no color. Three-quarters of the atmosphere of the Earth is composed of nitrogen (more than oxygen, in fact). This is one of the diatomic elements (it appears in nature as N_2). Nitrogen is commonly found in nitrates and in ammonia which are both utilized in the creation of fertilizers. Plants and animals are unable to metabolize nitrogen by themselves and, instead, rely on "nitrogen-fixing" bacteria to metabolize the nitrogen for their use.

8. **OXYGEN (O/8)**: Oxygen is a flammable, odorless, colorless gas which reacts readily with nearly every element in order to form "oxides" (molecules with oxygen). This is another diatomic molecule, appearing in nature as O_2. Oxygen makes up around one-fifth of the atmosphere of the Earth. The ozone layer of the Earth is comprised of O_3 molecules.

9. **FLUORINE (F/9):** Fluorine is usually found in nature as an extremely reactive pale yellow gas. It occurs as a diatomic molecule (F_2). Fluoride is an anionic form of fluorine which is used in making toothpaste and helps to strengthen tooth enamel.

10. **NEON (NE/10):** Neon is found in nature as a nonflammable, nonreactive, odorless and clear gas in nature. It is commonly utilized in the creation of neon signs, but there are often other gasses which are also used for that purpose as well.

11. **SODIUM (NA/11):** When found in its elemental state, sodium is a soft, yet solid, shiny metal that is very reactive. When mixed with water, for instance, it is able to break the bonds between the hydrogen and oxygen in the water molecule with enough heat and force that the hydrogen will combust. Sodium forms +1 cations easily, and it is usually found in this state when it is found in nature. Sodium is one of the most common elements you will find, usually bonded with others. For instance: table salt is NaCl, a compound comprised of sodium and chloride.

12. **MAGNESIUM (MG/12):** This is a shiny and solid metal which reacts with water when found at room temperature. When ignited, magnesium burns brightly. It is often used in match heads, some striker strips for ignition, and in other products used for combustion. It is also a major constituent part of the chlorophyll molecules which are used by plants during the process of photosynthesis.

13. **ALUMINUM (AL/13):** This is a lightweight and shiny metal, solid at room temperature, which is used in a variety of applications. It is used in everything from soda cans to the siding of houses. It forms an oxide coating which can prevent further reactions, which is partially what makes it so useful. It is used in areas where rust would be a concern. Before a standard process was found to isolate elemental aluminum, it was one of the rarest metals. So rare, in fact, that as a sign of the dominance of the United States, the top of the Washington Monument is capped with aluminum.

14. **SILICON (SI/14):** Silicon is a solid semimetal which is used to create a variety of objects, including quartz, glass, and sand, among other things. One of the largest modern applications for elemental silicon is the creation of computer chips. It is also utilized in many building applications, including caulking, where it is turned into a polymer known as silicone.

15. **PHOSPHOROUS (P/15):** This is a solid element at room temperature and occurs in three primary colors: black, red, and white. Red and white phosphorous are the most commonly found forms and are usually found as P_4 when they are found in nature. This is an extremely reactive element and usually has to be stored under special conditions (underwater, typically) so that it does not react with the oxygen found in the air and ignite. Fertilizers often use phosphorous compounds.

16. **SULFUR (S/16):** This is a yellow solid at room temperature. It is very brittle. When found in nature, it is found in the form S_8. It is a major constituent part of sulfuric acid, and its primary industrial application is in the creation of that acid. Sulfur emissions from industrial burning (such as coal power plants) can result in the eventual formation of H_2SO_4 (sulfuric acid) in the atmosphere. When this is brought down with rain, it is referred to as *acid rain*.

17. **CHLORINE (CL/17):** Chlorine is an element everyone is familiar with. Elementally, it is a yellow gas with a very choking odor. It is diatomic in nature (Cl_2). It forms anions, –1, very readily and, as a result, it is often utilized in substitution reactions. This is a major part of the brine which is found in the ocean. It is also used as a way to sterilize surfaces, pools, and other things. When released into the atmosphere, it contributes to the destruction of the ozone layer.

18. **ARGON (AR/18):** A clear and odorless gas, argon is nonreactive. It makes up a very small portion (1 percent) of the atmosphere. It is usually used in light bulbs, where it helps to cover filaments. It is also used in some industrial applications, such as welding when there is a need to avoid reactions.

19. **POTASSIUM (K/19):** This is a soft and solid metal which readily reacts when it is found in its elemental state. Potassium can react with water in a more violent fashion than even sodium is capable of doing. Only the +1 cationic form of potassium is found in nature. This is an extremely important element, playing a role in fertilizers which assist in plant growth and in the sodium/potassium pumps which help to assist in muscle contraction.

20. **CALCIUM (CA/20):** Calcium is a metal which reacts with water in a way similar to the way magnesium does. It is not a strong enough reaction to cause an ignition, however. Calcium is usually found as a cation (+2) in nature. It is the major element making up both skeletal

structures and tooth enamel. It is also used in a number of basic solutions, such as lime and calcium bicarbonate.

One thing to keep in mind when you look at the list and, in fact, when you look at the periodic table in general, is the fact that a given element and its symbol are not always obviously connected. You can look at potassium, for example, which starts with a P and has K as its symbol.

When you look at this list, you will probably realize that there are some properties which are unique to specific groups (columns). This is another benefit of the periodic table: substances which have similar properties will be grouped together. Below is a table of the most important groups on the periodic table of the elements.

Table 9.7 . Significant periodic table groups

Group	Elements	Characteristics
1A; alkali metals	Li, Na, K, Rb, Cs, Fr	form salts that can be dissolved in water; readily react to make +1 ions
2A; alkaline earth metals	Be, Mg, Ca, Sr, Ba, Ra	readily react to make 2+ ions
5A; pnictogens	N, P, As, Sb, Bi	readily react to form −3 ions
6A; chalcogens	O, S, Se, Te, Po	readily react to form −2 ions
7A; halogens	F, Cl, Br, I, At	readily react to form −1 ions
8A; noble gasses	He, Ne, Ar, Kr, Xe, Rn	none are particularly reactive; do not form compounds or ions regularly; outermost electron shells are already filled

Out of the twenty important elements that have already been outlined, six of them are even more important. Those elements are carbon, hydrogen, nitrogen, phosphorous, sulfur, and oxygen (CHNOPS is a common acronym for these six elements). These elements are the most important for life.

Organic Chemistry

Organic chemistry is one of the largest and most widely studied branches of chemistry. It is the primary bridge between biology and chemistry as well. Organic chemistry is the study of molecules based on carbon. Carbon is able to attach to many other carbon atoms in order to form longer, chained, molecules, so the variety of possible molecules is extremely large. Some examples of things studied by organic chemists are nucleic acids, amino acids, oils, tissues, proteins, plastics, and a slew of others (far too many to name).

The simplest carbon-based molecules (organic compounds) are named based on the number of carbon atoms that are found in a single chain. Below is a table of the most common prefixes used in organic chemistry.

Table 9.8. Common prefixes

Prefix	Carbon atoms in chain
meth-	1
eth-	2
prop-	3
but-	4
pent-	5
hex-	6
hept-	7
oct-	8
non-	9
dec-	10

Organic compounds are what forms the basis of all life on Earth. They are also the primary driver behind the vast majority of experiments conducted by chemists. The way that carbon is able to bond, forming four total bonds, allows for such a wide array of diverse carbon-based compounds that all the different proteins, pharmaceuticals, and types of life on Earth are able to stem from them.

Metals

Most elements found on the periodic table are metals. Many metals are found in nature in an oxidized state (combined with either sulfur or oxygen). Some metals, like the ones commonly found in mines (gold, silver, copper, etc.) can be found in an elemental state, however. Metals are capable of forming ALLOYS. Alloys are mixtures of two or more metals. Allows are solid. You may be familiar with the term AMALGAM if you have been to the dentist a few times. This is a mixture of mercury and another metal. A given amalgam can be either liquid or solid. The state is dependent on the amount of mercury which is present in the mixture.

Metals all share a few common properties, all of which are outlined below:

- They are usually silver in color.
- They are shiny.
- They readily conduct both electricity and heat.
- They can be cut into thin sheets (sectile).
- They can be drawn into thin wires (ductile).
- They can be hammered into thin sheets (malleable).
- They are solid at room temperature.

Metals will usually have all of these properties at differing levels, however some don't follow all of the rules. Mercury, for example, is liquid at room temperature, not solid.

Energy and Radioactivity

Energy is a property which objects can have that can be transferred to other objects or can be converted into another form. In the same way that the law of conservation of mass states that mass cannot be created or destroyed, the law of conservation of energy states that energy cannot be created or destroyed. It can, however, be converted into different forms. Energy can be present as a reactant in a reaction or as the product. It can also be utilized to make the reaction happen as well.

Two types of energy exist: potential and kinetic energy. **POTENTIAL ENERGY** is energy that is being stored. It has the "potential" to be used and to do work. This is usually dependent on either the types of chemical bonds that are present or, in a more real world example, the distance an object is from the ground. A rollercoaster at the top of its track is going to have a lot of potential energy, for example.

KINETIC ENERGY is the type of energy being used for motion. If something is moving slowly, it has a low kinetic energy. The faster that object starts to move, the higher its kinetic energy is going to go. This type of energy is often shown in temperature. When things are hot, their atoms and molecules are moving faster than when they are cold.

The energy which is being stored inside the nucleus of a given atom is a form of potential energy. This is the type of energy that is being used in radiation treatments, in nuclear power plants, and in the creation of nuclear bombs. When a nucleus which is unstable converts (through decomposition) to a stable nucleus, that potential energy is released.

The following are a few definitions which can help you get through any questions about radiation:

- **RADIOACTIVITY** occurs when particles are released from a nucleus as a result of instability in the nucleus.
- **GAMMA RAYS** are high energy particles of light.
- An **ALPHA PARTICLE** is two neutrons and two protons.
- A **BETA PARTICLE** is an electron.
- A **NEUTRON** is a neutrally charged subatomic particle with a weight similar to that of a proton.

Physics

Physics is the study of matter and the motion of matter through space and time (surely you can see how this subject overlaps a bit with chemistry). Some concepts related to this include energy and force.

Physics is broadly concerned with the natural laws which govern the way that the universe as a whole is going to behave.

Motion

Motion is what happens when something moves from one place to another place. Three types of motion exist: translational motion, rotational motion, and vibrational motion.

Put bluntly, **TRANSLATIONAL MOTION** is motion happening in a straight line (on a linear axis). It is defined by a change in the position of an object over a given period of time and by movement relative to a fixed reference point. **ROTATIONAL MOTION** is a motion that is happening around an axis, and lastly **VIBRATIONAL MOTION** is motion which is occurring around a single, fixed, point.

Physicists use a variety of terms in order to quantitatively define motion, including speed, velocity, acceleration, scalar, and vector.

Generally, **SPEED** is going to be a measure of how quickly something is moving by dividing the distance it has traveled by the time is has taken to do so:

$$speed = \frac{distance}{time}$$

VELOCITY is also a way to measure how quickly something is moving. However, as opposed to speed, velocity needs a third piece of information: direction. Velocity is known as a vector quantity (more on this later).

$$velocity = \frac{displacement}{time}$$

ACCELERATION is a change in velocity over a given period of time. When that velocity increases, acceleration is the term which is used. When it decreases, the term deceleration is used. This is also a vector quantity. It is positive when it is occurring, in the same way, a given object is moving and negative when it is moving the opposite way.

$$acceleration = \frac{change\ in\ velocity}{time}$$

A **SCALAR** quantity is one in which direction does not make a difference. A **VECTOR** quantity is one where the direction traveled does make a difference.

Graphs are used in order to quickly show the way an object is moving. Generally, two types of graphs are going to be used. The first is a **POSITION-TIME GRAPH**. This shows how the position of an object is changing over time. The velocity on this type of graph is going to be equal to the slope of the line on the graph.

The second type of graph is a **VELOCITY-TIME GRAPH**. This type of graph is meant to quickly show how the velocity of a given object is changing over a period of time. The slow on this type of graph is going to be the acceleration. This can also be used to figure out the distance being traveled by an object using the area underneath the line on the graph.

LINEAR MOTION is motion which is occurring in a single dimension, along the *x*-coordinates of a graph. Motion on the *y*-coordinate is known as **MOTION IN A VERTICAL PLANE**. Usually, the acceleration along either of these is going to be constant. Since that is the case, the acceleration itself can be considered a constant. During this situation, motion can be described by the **EQUATIONS OF KINEMATICS**.

KINEMATICS: IMPORTANT EQUATIONS

These equations are the main equations which are used in order to calculate acceleration and velocity. To get started, you need to understand the five primary variables utilized in these equations:

- x = displacement (distance traveled)
- a = acceleration
- t = time
- v = final velocity
- v_0 = initial velocity

The relationships between these five variables gives you the following four equations:

$$v = v_0 + at$$

$$x = \left(\frac{1}{2}\right)(v_0 + v) \times t$$

$$x = v_0 t + \left(\frac{1}{2}\right) \times at^2$$

$$v^2 = v_0^2 + 2ax$$

These equations all have four out of the five variables that were defined at the outset. That is the method through which you can find a missing variable. Which four are on the question you are looking at? Pick the right equation for that situation, plug in the relevant numbers, then solve for the variable you need.

Now that these definitions and equations have been outlined, it is time to talk about motion which is occurring in a vertical plane. One thing to note at the outset: The force of gravity causes an acceleration (on Earth) of 9.81 m/s². When you use the kinematic equations to describe an object moving in freefall, you will utilize g (gravity) instead of acceleration, in the same way, you will use y instead of x.

THE LAWS OF MOTION: ISAAC NEWTON

Isaac Newton was one of the most profound physicists ever to live. His ideas and findings are still used to this day in order to help describe motion. Before outlining his three laws of motion, the term force needs to be defined. A **FORCE** is a push or pull which can cause a resting object to move or cause a change in the velocity of an object that is already in motion. The three laws of motion (in order) are as follows:

1. An object at rest will stay at rest, and an object that is in motion will remain in motion at constant velocity unless a force acts upon it.

2. The acceleration of an object is directly proportional to the force that is acting on it and is inversely proportional to the mass of the object.

3. If an object exerts a force on another object, the second object will exert an equal force back on the first object, but in the opposite direction.

Newton's first law describes a concept known as INERTIA and is commonly known as the law of inertia. Inertia is the tendency of a given object to resist any change to its motion.

The second law of motion is meant to describe how forces act on objects. The general equation that is derived from the second law is the following, where m = mass and F = force::

$$F = ma$$

Force is another vector quantity. It has a size and a direction. Like velocity, is can be positive or negative, depending on the direction it is being applied.

Energy

Energy and work are terms which go hand in hand. When scientists talk about what mass a given object has, they are also referring to the amount of energy it has. Energy has two main types, kinetic and potential energy, as discussed previously.

Both types of energy will change when an object is either doing work or when work is being done to that object. WORK can be defined as a transfer of energy to an object when that object is put into motion because of a force acting on it. The following formula can be used to define work, where F = force (newtons), W = work (joules), and d = distance (meters):

$$W = Fd$$

So what is energy, anyway? Energy is the capacity to do work. If you pick up a ball and lift it over your head, you are doing work and supplying that ball with potential energy. The force that you are working against is the force of gravity, in that case. The work that is being done against the force of gravity is known as GRAVITATIONAL POTENTIAL ENERGY. The following equation can be used in order to calculate the gravitational potential energy of an object, where PE = potential energy (joules), m = mass (kilograms), g = acceleration due to gravity, and h = height:

$$PE = mgh$$

When an object is at a height, it has potential energy. When it begins to fall, however, it is going to be accelerated by the force of gravity and, thus, lose some of the gravitational potential energy that it has. Because of the law of conservation of energy (as discussed previously), energy is simply being converted from potential energy to kinetic energy. The decrease in one type of energy is immediately associated with an increase in the other type. Kinetic energy of moving objects also has an equation associated with it, where KE = kinetic energy (joules), m = mass (kilograms), and v = velocity (meters per second):

$$KE = (\tfrac{1}{2})mv^2$$

Many types of objects utilize the above principles in order to change potential energy to kinetic energy and vice versa (and types of energy between one another). Here are a few of them:

- **SOLAR CELLS** convert light energy into electrical energy.
- **BATTERIES** convert chemical energy into electrical energy.
- **LAMPS** convert electrical energy into heat and light.
- **HEATERS** convert electrical energy into heat.
- **FIRE** converts chemical energy into heat and light.

There are, of course, many others. Dams, generators, car motors, alternators, etc. All of these things utilize one form of energy in order to do a conversion. Many of them are essential to the way people live in the modern age.

There is one other term which needs to be discussed here: power. **POWER** is the rate in which something is able to convert energy from kinetic energy to potential energy (or the other way around) using work. Here is the equation used to determine power, where P = power (watts), W = work (joules), and t = time (seconds):

$$P = \frac{w}{t}$$

Fluid Mechanics

Fluid mechanics is the study of fluids and the way that they react to physical pressures. A **FLUID** is a substance which is going to offer minimal resistance to changes in the shape that it has when pressure is applied. Both liquids and gasses can be considered fluids. Solids, however, are not. Many properties help to define fluids as a whole, but the most important is the capability of fluids to exert pressure on things.

What is pressure? Pressure is a force which is being exerted per unit of area. The general equation for defining pressure, where P = pressure (pascals), F = force (newtons), and A = area (square meters), is as follows:

$$P = \frac{F}{A}$$

The way that fluids are able to exert pressure on things is explained by something known as the **KINETIC MOLECULAR THEORY**. This theory states that particles making up various fluids are continuously undergoing random motion. The particles will, as a result, constantly be colliding both with each other and with surfaces that they make contact with. When these particles make that contact, a force is exerted. The combined force of all of those collisions is pressure.

There are a three primary principles which govern the way that fluids work. The first one of these principles is **ARCHIMEDES' PRINCIPLE**. This principle states that if an object is covered by a fluid, it is going to be buoyed up by a force that is equal to the weight of the fluid that is being displaced by that object. The following formula can be used to determine the size of that force, where F = force (newtons), ρ = density (fluid), V = volume (fluid), and g = acceleration (gravity):

$$F = \rho V g$$

Any object which is being immersed in a given fluid is going to either float or sink. Which one it does depends on the weight of the object being immersed relative to the force that is being exerted on it by the fluid it is immersed in. If the force is less than the weight of the object, then the object is going to sink. If the two values are equal, the object can float in the liquid at any given depth. If the force is greater than the weight, the object will float on the surface of the liquid. This principle is frequently used to determine the weight of an object by measuring its fluid displacement in water and, in fact, that was the reason that it was originally created.

The second principle is Bernoulli's Principle, and the third is Pascal's Principle.

Electricity

Electricity is the phenomenon which is associated with the presence of an electrical charge. A wide range of things are associated with electricity including electric current, lightning, static electricity, and electromagnetism. Electricity is primarily caused by a flow of electrons.

The following are a few definitions that you should get to know:

- **ELECTRICITY** is the flow of electrical energy from one place to another (usually from an electrical power source to whatever is using it).

- **ELECTRIC CURRENT** is the method by which electrical energy is transported; it is the flow of electrons. Current is measured in amperes.

- The **LOAD** is the component or part of a given circuit which is consuming the electrical power in an electrical system.

- An **ELECTRON** is a negatively charged subatomic particle.

- A **CIRCUIT** is a system of metal wires capable of conducting electricity which then provide a pathway for current to flow.

- A **CONDUCTOR** is any substance that allows the flow of an electric current.

- An INSULATOR is any substance which does not allow for the flow of an electric current.

- VOLTAGE is the potential difference between two ends of a conductor. This is what allows for the flow of the electric current. Voltage is measured in volts.

- RESISTANCE is the strength to which a material "resists" the flow of electrical current. Resistance is measured in ohms.

- RESISTIVITY is a property of materials which helps to determine the amount of resistance that those materials have to the flow of electrical current.

- ELECTROMAGNETISM is the property of moving electrical charges to produce a magnetic field. Likewise, magnetic fields which are changing are able to generate currents of electricity.

Some of these terms are going to go together. The higher the voltage, the higher the current. The lower the voltage, the lower the current. Insulators and conductors are often used in conjunction with one another in order to create methods of carrying electrical current over long distances without losing any of the energy. Think about the rubber (insulator) wrapped around the copper (conductor) wires in your house, for instance.

Ohm's law is the equation and theory which govern the way voltage, resistance, and current relate to each other. It states that the current between two points in a conductor is going to be proportional to the potential difference (voltage) between those points, where I = current (in amperes), V = voltage (in volts), and R = resistance (in ohms):

$$I = \frac{V}{R}$$

You can use this equation in order to solve for any one of the three variables inside of it, provided you know the other two. This is a powerful equation which is used in many fields, including one that is covered in another the electronics subtest.

Sound

Sound is a type of vibration which moves through a given medium as a mechanically derived wave of both displacement and pressure. The medium through which sound waves travel could be air, water, or another fluid-like substance. The sound which is perceived changes depending on the type of medium through which the waves are traveling.

Sound waves are pressure variations which are being carried through matter. There are two types of pressure variations: RAREFACTIONS, areas which have low pressure, and COMPRESSIONS, areas which have high pressure.

These two pressure variations are created when the sound wave makes the molecules of air (or whatever medium they are traveling through) collide with each other and, thus, make the pressure variations move away from the sound source. One of the logical conclusions that you may be able to draw from this (or perhaps you already knew) is that sound cannot travel in a vacuum. Because vacuums inherently have no particles to be moved, the sound waves and pressure variations cannot propagate within them.

How quickly sound is able to travel depends on temperature as well. For example: in the air, at room temperature, sound waves can travel about 343 meters per second. Another variable which can affect the speed at which sound travels is pressure.

It is worth noting that sound can travel through mediums besides air as well. It can travel through solids and liquids. The speed at which sound travels through solids and liquids, however, is faster than it is through the air. When sound waves hit solid surfaces, they can be reflected back. You know these reflections as ECHOES. How long an echo takes to travel back is going to depend on how close the surface it is reflecting off of is to the original source of the sound wave. Some animals use echoes in order to figure out where they are, spatially, and to help locate things. Bats do this, as do dolphins. This is also the method through which sonar works on marine vehicles.

The combination of compressions and rarefactions is also called the wavelength. The number of waves coming in a single second is known as the FREQUENCY of the sound. The term PITCH is sometimes used in order to describe the same thing. If the source making the sound is moving, then someone listening will hear the sound changing into different frequencies as a result. When moving away, the frequency goes down, when moving closer, the frequency goes up. This trend of changing frequencies is what scientists call the DOPPLER EFFECT and is commonly used in applications such as ultrasound, sonar systems, and radar detectors.

Earth Science and Astronomy

Earth science is a term which encompasses all of the various fields of science that deal with the Earth itself. This is one of the oldest historical sciences. Included in it are the studies of the stars (astronomy), the oceans (oceanography), the Earth itself (geology), and the weather (meteorology). These are not, of course, the only subjects which are studied in Earth Science. Everything from the atmosphere and the hydrosphere to the biosphere is all included in this catch-all term. Further, other general science principles are utilized in the study of Earth science, including mathematics, physics, biology, and chemistry. The environment and ecology are also a primary focus of some scientists studying Earth science.

On the AFOQT, you will encounter questions in astronomy, oceanography, geology, and meteorology.

Astronomy

Astronomy is a science which is concerned with the study of celestial objects and everything that occurs outside of the Earth's atmosphere. This includes everything from planets and stars to, at times, the nature of the universe itself (though that is usually left to a branch of science known as physical cosmology).

Below is a brief list of some vocabulary you might encounter in the astronomy portion of the general science subtest:

- A **PLANET** is an astronomical object which is large enough to have its own gravity but not large enough to undergo thermonuclear fusion. It is also large enough to have cleared the local area of other celestial objects smaller than it.

- A **STAR** is a luminous plasma ball which is being held together by nothing but its own gravity. The Sun is a star, as are the vast majority of lights seen in the night sky.

- A **SOLAR SYSTEM** is a star and all of the objects which are orbiting that star. The objects do not necessarily have to be planets.

- A **GALAXY** is a system of stars, interstellar gas, dust, dark matter, and stellar remnants which are bound together by an immense gravity field.

- The **SUN** is the star found at the center of the solar system. It is also the primary source of energy for the planets which orbit it (including the Earth).

- An **ELLIPTICAL PATH** (or elliptical orbit) is a type of orbit around an object which is roughly egg-shaped, rather than being perfectly spherical.

- **HELIOCENTRIC** is a term that describes a system which has a sun as its center, such as our solar system.

- A **ROTATION** is the circular movement of a given object around a single point (usually the center of the object).

- **REVOLUTION** might be considered another term for orbit, astronomically speaking. This term is used when one object moves around another one.

Here are some facts about the Earth:

- The rotation and revolution of the Earth, along with its tilt, are what lead to the day and night cycle.

- The axial tilt toward the sun along with the revolution are what lead to the years and seasons.

- Local weather is usually a result of the amount of solar radiation which is able to enter the atmosphere.

- The axis of rotation of the Earth is 23.5 degrees.

- The Earth rotates along its axis while, at the same time, moving counterclockwise around the sun.

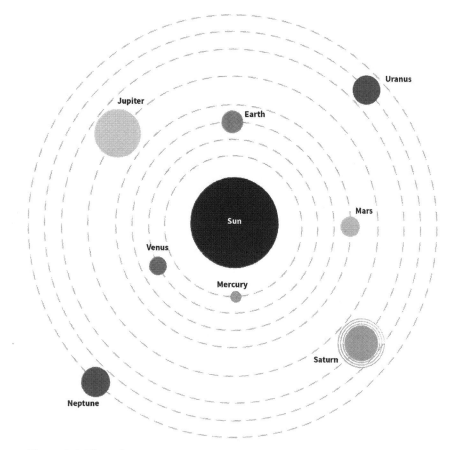

Figure 9.6. The solar system

The sun of the solar system that the Earth is in puts out such an immense gravitational field that it is able to hold all of the planets in their orbits around it. That solar system is also heliocentric, meaning that the sun is the center of the solar system. There are eight planets in the solar system which follow elliptical orbits around the sun. The output of electromagnetic radiation from the sun is what provides the energy from which life on Earth is derived.

MERCURY is the planet which is closest to the sun. The size of this planet is not much larger than the moon of the Earth. On the day side, the temperatures reach around 840 degrees Fahrenheit while, on the night side, temperatures can drop to very far below freezing. There is nearly no atmosphere on mercury and, thus, there is nothing to protect it from impacts from meteors.

VENUS is the second planet from the sun. It is even hotter than Mercury, and the atmosphere is toxic due to a greenhouse effect (which also traps heat). The pressure on the surface of the planet is immense. Venus has a very slow rotation around its axis and spins in the opposite direction than most of the planets of the solar system. Venus is one of the brightest objects in the sky, outside of the moon and the sun.

EARTH is the third planet from the sun. This is a water dominant planet, having around two-thirds of the surface covered by water. This is the only known planet which has life on it. The atmosphere consists primarily of nitrogen and oxygen. The term *day* and *year* are based on the rotation of the Earth around its axis and the revolution of the Earth around the sun, respectively.

MARS is the fourth planet from the sun and is a cold and dry planet. The dust which makes up the surface of the planet is a form of iron oxide, which is also what gives the planet its red color. The topography of Mars is very similar to that of Earth. Ice is located in some locations on Mars. Additionally, though the atmosphere is currently too thin for water to exist on the surface in liquid form, it is theorized that, in the past, water was abundant on the planet.

JUPITER is the largest planet in the solar system. It is the fifth planet from the sun. The planet itself is primarily gaseous, being composed of hydrogen and helium (along with others, in smaller amounts). The Great Red Spot is an enormous storm which has been ongoing on the planet for hundreds of years. The planet has dozens of moons and a very strong magnetic field.

The sixth planet from the sun, SATURN is primarily known because of the rings which orbit it. The planet is primarily gaseous, being comprised of helium and hydrogen (predominantly). The planet has multiple moons.

URANUS is the seventh planet from the sun. The equator of Uranus is at a right angle to the orbit of the planet, leading to it appearing to be on its side. This planet is about the same size as Neptune. Uranus has faint rings, multiple moons, methane, and a blueish-green tint.

Eighth planet from the sun, NEPTUNE is characterized by its cold temperature and its very strong winds. The core of Neptune is rocky. Neptune is about seventeen times the size of Earth. The planet was originally discovered using the power of math, by theorizing about the irregular orbit of Uranus being the result of a gravitational pull.

In recent years, PLUTO, formerly the ninth planet from the sun, has been reclassified from a true planet to a dwarf planet. The AFOQT is unlikely to ask any tricky questions about Pluto, especially considering it has only recently changed classifications. Pluto is smaller than the moon of the Earth, and it has an orbit near the outer edge of the local solar system. Its orbit around the sun takes around 248 years on Earth. The planet itself is rocky and very, very cold. Its atmosphere, what little there is, is extremely thin.

Oceanography

About 71 percent of the surface of the Earth itself is water. Relatively speaking, the water layer is not very thick in most places. The water contains minerals and salts which have been dissolved, and the vast majority of the water is held in four ocean basins. Seas, porous rocks, ice caps, lakes, and rivers contain the rest of the water on Earth. From smallest to largest, the oceans are the Arctic Ocean, the Indian Ocean, the Atlantic Ocean, and the Pacific Ocean.

The oceans play a very large role in the maintenance of the environment of the Earth. Things such as the shape of the Earth, the axial tilt, the way it rotates, and the relative distance from the sun affect the heating across the oceans. The water allows that heat to be distributed around the planet. That is also what helps to create ocean currents. Oceans are able to absorb and release heat and, thus, they help to regulate both the climate and the weather.

Ocean basin is a catch-all term for the land which is under the ocean. Generally speaking, these are comprised of basaltic rock. Some areas of the basin have a large amount of both seismic and crustal activity, however, and they are the areas which are responsible for the activity studied in plate tectonics.

Oceanography itself is a pretty broad topic and is a term used to describe the study of the ocean as well as oceanic ecosystems, currents, fluid dynamics, plate tectonics, and marine organisms. Typically, oceanography is going to fall into a multidisciplinary area. One of the major areas of study in oceanography, currently, is the study of the acidification of the ocean, which is a term used to describe the way that the pH of the ocean is decreasing due to carbon dioxide emissions into the atmosphere of the Earth.

Geology

The study of geology is the study of the Earth itself. The Earth, as you may know, is a planet which is revolving around the sun. The Earth is not uniform in shape (it is not a perfect sphere). It has a number of layers and a disparate topography. Among the layers of the Earth are the crust, the asthenosphere, the mantle, and the core. The thinnest of those layers is the crust, which is also the outermost layer (and the layer that humanity calls home). The crust of the Earth is about eight thousand miles around (in diameter). Together with the topmost layer of the mantle (the most solid part of the mantle), the two are known as the lithosphere.

GO ON

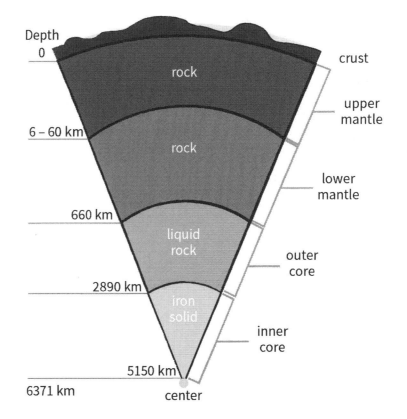

Figure 9.7. Layers of the Earth

The way that the interior of the Earth functions is a sub-discipline of geology known as seismology, which is the study of the seismic activity of the Earth. Seismic activity, for all intents and purposes, is another term for earthquakes.

The crust is not uniform at all and, in fact, has a unique topology. Think about the crust and what you know about it. Valleys, mountains, plains, oceans, etc. These are all variations in the height and thickness of the crust. The portion of the crust which forms continents is primarily granitic rock. On the other hand, the part of the crust which is located underneath the oceans is primarily comprised of basaltic rock. These two types of crust are then broken up into tectonic plates which move over the asthenosphere.

The movement of the tectonic plates on each other is known as (big surprise) plate tectonics. This is what explains many natural phenomena that happen on the Earth, including volcanoes, how mountains are made (and destroyed), the way the sea floor changes, trenches in the ocean, and earthquakes. The crust is usually only stable for a small time frame, geologically speaking. Earthquakes always occur, volcanoes become active and go dormant, and many other things are ever-changing on the crust of the Earth.

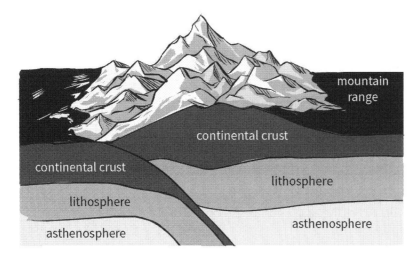

Figure 9.8. The Earth's crust

Rocks on the Earth are also changing. Generally, rocks are made up of a combination of different inorganic crystalline substances known as minerals. Each type of mineral will have a specific chemical makeup and will have properties that are unique to them. Here are some of the most common minerals that you might encounter:

- BAUXITE is a type of rock composed primarily of aluminum oxides which have been hydrated.

- QUARTZ is the most abundant mineral in the crust of the Earth. This is the most common and simple form of all silicates. It is an oxide of silicon.

- TALC is a common and soft mineral which can be scratched with a fingernail.

- PYRITE is a mineral comprised of iron and sulfur. Pyrite is commonly known as "fool's gold" because of its resemblance to its namesake.

- GRAPHITE is a form of carbon. You may recognize this as being used in pencils and some other commercial applications.

Also included in this list would be all of the types of precious stones you can think of: emerald, ruby, opal, diamond, sapphire, etc.

Minerals comprise the types of rock that make up the Earth as well. The most common types of rocks are igneous, sedimentary, and metamorhic. IGNEOUS ROCKS have been formed through the melting and cooling of minerals within the mantle beneath the surface of the Earth. These can surface, initially, as lava. Some common examples of igneous rocks are granite, basalt, and solid volcanic lava.

SEDIMENTARY ROCKS have been created by the deposit of their composite materials onto the crust of the Earth and inside the bodies of water that help make up its surface.

METAMORPHIC ROCKS are rocks that already exist, but have gone through a transformation. When rock is put under immense heat and pressure, changes can occur within it. Some of the examples of this type of rock would be slate, gneiss, and marble. The changes occurring within the original rock can be both physical and chemical in nature.

There are two layers which are not layers of the planet itself, but which the crust is nevertheless in contact with—atmosphere and hydrosphere. ATMOSPHERE is a term for the layer of gasses which surround planets (or any large body with a significant amount of mass) which are being held by the gravity of those planets. The atmosphere of the Earth is primarily nitrogen and oxygen and has multiple layers.

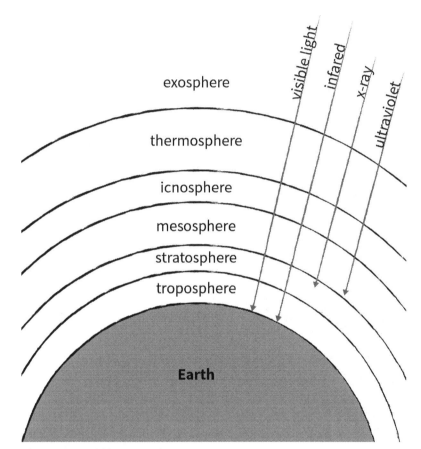

Figure 9.9. Earth's atmosphere

The HYDROSPHERE is the term which is used for the collective water of the Earth. This includes everything from oceans to lakes, ponds, and rivers. On the Earth, the hydrosphere is about 70 percent of the surface.

The way that these two layers meet with the crust results in an energy exchange which manifests itself through three processes: weathering, erosion, and deposition.

WEATHERING is the process of rocks and soil breaking down through contact with the atmosphere and waters of the Earth. This occurs without movement. It is important to note that this is not the same thing as erosion.

EROSION is the process of rocks and soil breaking down and being moved and deposited somewhere else through nature processes such as the flowing of water, wind, and storms.

DEPOSITION is a process through which soil and rocks are added to a mass through transport as a result of erosion and a loss of kinetic energy. Deposition would be rocks breaking off of a mountain because of a hard storm and then "depositing" down at the bottom of the mountain when they no longer have enough kinetic energy to move.

Through these processes, the crust of the Earth is always being changed. It is changing and being changed by natural forces.

Meteorology

METEOROLOGY is the study of the atmosphere, weather, and climate of the Earth. The atmosphere of the Earth is meant to help reflect, refract, and absorb the light energy being emitted from the sun. The process of these activities taking place is what leads to the energy flow that causes the weather and climate of the Earth.

WEATHER is a temporary, day to day, type of atmospheric condition. Think about rainstorms, sunny days, cold days, or snow. All of these are, generally, temporary conditions. They are not long-term, geologically speaking.

A CLIMATE is the longer term version of weather. Think about how California is generally hot and dry. How Florida is hot and wet. The northern states are cold. Jungles are hot and wet. Deserts are dry. All of these are types of climates. They are not temporary like normal weather would be.

The atmosphere of the Earth consist of many elements, but it is primarily composed of nitrogen and oxygen. About 1 percent of the atmosphere is made up of other gasses, such as carbon dioxide, ozone, and argon. As far as nitrogen is concerned, it makes up around 78 percent of the atmosphere. Oxygen makes up 21 percent.

There are multiple layers of the atmosphere, all of which are separated from each other by pauses (which have the largest variation in characteristics): exosphere, thermosphere, mesosphere, strasphere, and troposphere.

The EXOSPHERE is the farthest out part of the atmosphere. This is where satellites are orbiting the planet and where molecules have the potential to escape into space itself. The very bottom of the exosphere is known as the thermopause. The thermopause is about 375 miles above the surface of the Earth. The outermost boundary of the exosphere is about 6,200 miles above the surface of the Earth.

The THERMOSPHERE is the next layer when coming toward Earth from space. It is just inside of the exosphere, separated from it by the thermopause. This layer is between 53 and 375 miles above the surface of the Earth. This layer is known, colloquially, as the upper atmosphere. The gasses in this layer are very thin, but they become more and denser the closer you get to the Earth. This is the layer which absorbs the energy coming in from the sun (particularly ultraviolet radiation and x-ray radiation). That energy is what leads to the high temperature. The top of this layer is around –184° Fahrenheit while the bottom is around 3,600° Fahrenheit.

Between thirty-one and fifty-three miles above the surface of the Earth, lies a denser layer of atmosphere and gasses called the MESO-SPHERE. The temperatures in this layer are around 5° Fahrenheit. Gasses here are usually thick enough to stop most small meteors that enter the atmosphere of the Earth, causing them to burn up. This layer, along with the stratosphere, are collectively known as the middle atmosphere. The boundary between the two layers is called the stratopause.

The STRATOSPHERE is from thirty-one miles above the surface of the Earth to around eight miles above the surface of the Earth, give or take four miles. About 19 percent of the gasses in the atmosphere are contained in the stratosphere, which has a very low water content. The temperature of this layer increases along with the height. The heat is a byproduct of the creation of ozone in this layer. The barrier between this layer and the troposphere is the tropopause.

The TROPOSPHERE is the lowest layer. This is where weather takes place. It goes from the surface of the Earth to between four and twelve miles above the surface. The exact height of this layer varies. The density of gas in the troposphere decreases with the height, and the air becomes thinner (which is why mountaintops have thin air).

The layer which humans live in is known as the troposphere, which is also the layer in which weather takes place. Some of the variables which are governed by weather conditions include the weight of the air (barometric pressure), air temperature, humidity, air velocity, clouds, and the levels of precipitation. Instruments are commonly used to help determine the relative levels of all of these. Below are some common weather measurement instruments:

- A THERMOMETER is a tool which is used to measure the temperature of the air using either mercury or alcohol.
- A BAROMETER would be used to measure the pressure of the air. If the pressure is going up, calm weather is coming. If the pressure if going down, expect rain.
- A PSYCHROMETERS is a device used to measure relative humidity through the use of evaporation.
- An ANEMOMETER is an instrument which measures wind speed using a series of cups which catch the wind and then turn a dial.
- A RAIN GAUGE is used to determine how much rain has fallen in a given period of time.

- **WIND VANES** are weather instruments which are used to help determine the direction of the wind.
- A **HYGROMETER** is an instrument which is used to help measure the humidity of the air.
- **WEATHER MAPS** show the atmospheric conditions over geographical areas. You are likely familiar with these from your local news.
- A **COMPASS** is an instrument which is used primarily for navigation and can be used to help determine directions.
- **WEATHER BALLOONS** are commonly used to help measure the conditions of weather in the upper atmosphere.

Obviously you will probably not have all of these in your home. It is likely, however, that you know of most of them (even if you do not know their names). It is also likely that you have used a few (or seen them used) at some point in your life.

Most meteorologists will utilize readings from a weather map and a number of these tools in order to determine what air masses are moving around in the troposphere. Air masses would be defined as sections of air that has a uniform content of moisture and temperature. The way that these air masses move and interact with each other determine the weather and, thus, are what most meteorologists use when determining changes in the weather.

The climate of the Earth and the seasonal changes which regularly occur are a result of the interaction between local water sources, the way that the Earth is tilted toward the sun, the altitude of the location in question, and the latitude of the location in question. These events are, obviously, very complex. It is not an easy task to help determine how the climate will change over time.

Tips

Below are some tips which will help you get through the general science subtest of the AFOQT:

- Don't spend too much time on any single question. If you don't know it, make an educated guess and move on.
- Try to eliminate ridiculous answers immediately. Usually, two or possibly three of the answers for any given question will be so wrong you don't even need to evaluate them for any significant amount of time.
- It will not do you any good to try to learn everything there is to know about science right off the bat. Instead, study in bite sized chunks and review regularly. If you find yourself weak in one or more areas, then spend a little extra time on them.

- Creating acronyms and tongue twisters can immensely help when attempting to remember all of the facts and figures in this section.

Review

Basic Principles

This includes the metric system, conversion back and forth, basic temperature information, and other commonly used scientific jargon which is included in all of the other sections of the subtest (and a number of others).

Scientific Method

The scientific method is the means by which nearly all scientific discoveries are made. The steps are simple: observe, hypothesize, predict, experiment, and repeat.

Disciplines

The disciplines, broadly, which are included on the AFOQT general sciences subtest are life science, chemistry, physics, and earth science.

Life Science

Biology and the fields of science which pay close attention to the study of living organisms.

- CLASSIFICATION is the way that organisms are grouped together for description and cataloging.
- The THEORY OF EVOLUTION describes the way that organisms and systems can change over time as a result of the environment that they are found in.
- CELLS are the basic unit of life. This section talks about cells, organelles, and the way that cells interact with each other.
- RESPIRATION, FERMENTATION, PHOTOSYNTHESIS are the three methods through which cells are able to create energy.
- CELL DIVISION is the study of the replication of cells, both mitosis and meiosis are included.
- GENETICS is the study of genes and how they function. DNA to RNA to protein is the central dogma of modern genetics.
- ANATOMY is the study of the human body and how it functions, including the many parts of the body.
- ECOLOGY is the study of environments and the interaction of plants and animals within those environments.

Chemistry

The type of physical science which is primarily concerned with the structure, properties, and composition of matter itself.

- ATOMIC STRUCTURE means atoms, their composition, and the subatomic particles that make them up.
- ATOMS are the most basic form of elements; MOLECULES are combinations of one or more types of atoms, and COMPOUNDS are combinations of molecules.
- The PERIODIC TABLE is the best way to list all of the elements that exist along with basic information about each of them.
- REACTIONS AND EQUATIONS explain the way chemical reactions are written, balanced, and explained.
- SOLUTIONS, ACIDS, AND BASES are mixtures of chemicals; information about how hydrogen or hydroxide interacts inside mixtures.
- Information about the most important ELEMENTS that are found in nature.
- ORGANIC CHEMISTRY is the study of compounds and molecules which contain carbon.
- Information about METALS and the characteristics of common metals that are found in nature.
- ENERGY AND RADIOACTIVITY as it relates to chemicals and bonds between molecules and atoms.

Physics

A natural science which is concerned with the study of matter and how it moves through time and space.

- MOTION is the study of changes in position or time and the study of movement itself.
- KINEMATICS is a subdiscipline of mechanics which is concerned with the motion of objects without considering the cause of the motion.
- The LAWS OF MOTION as outlined by Isaac Newton. These define the way that objects and forces acting on them relate to each other.
- KINETIC AND POTENTIAL ENERGY as it relates to physics.
- FLUID MECHANICS is how fluids move and react to their environment.
- ELECTRICITY is a basic primer on the physical basis of electricity, including the interactions between subatomic particles which lead to its manifestation in the environment.
- What SOUND is, the way sound travels, and how sounds are amplified when they need to be.

Earth Science and Astronomy

Primarily the study of the Earth itself and the way that it relates to other bodies in the universe.

- **ASTRONOMY** is the study of the stars, the planets, and space.
- **OCEANOGRAPHY** is a multidisciplinary study of the oceans. This includes the currents, marine biology, plate tectonics, and other subjects involving the oceans.
- **GEOLOGY** is the study of the Earth itself, including the types of rocks it is composed of and the way that the rocks change over time.
- **METEOROLOGY** is a term used to describe the study of the local and macro-climates of the Earth, including the weather.

Practice Problems

1. The exosphere is the outermost layer of the atmosphere. How far above the surface of the Earth is the outermost boundary of the exosphere?

   1-A 6,200 miles

   1-B 200 kilometers

   1-C 100,000 miles

   1-D 5,000 miles

2. Sea water is composed primarily of water (H_2O) and _____.

   2-A fish

   2-B potassium chloride

   2-C gold

   2-D sodium chloride

3. This organelle is known as the powerhouse of the cell. It is where ATP is first created inside cells.

   3-A plasma membrane

   3-B mitochondria

   3-C nucleus

   3-D ribosome

4. In our local solar system, there are eight planets. What do these plants rotate around? What is the body at the center of the local solar system?

   4-A sun

   4-B moon

   4-C Earth

   4-D Venus

5. What is the term which describes interactions between communities of organisms and the environment that they live in?

   5-A foot system

   5-B community

   5-C ecosystem

   5-D environment

6. Toothpaste is a compound which is used to help prevent the decaying of teeth. What element is added to toothpaste in order to accomplish this?

   6-A fluoride

   6-B calcium

   6-C silver

   6-D amalgam

7. The Earth is made up of multiple layers. What is the outermost layer of the Earth itself?

   7-A core

   7-B crust

   7-C mantle

   7-D oceans

8. Which kingdom is the one in which ameba belong?

   8-A homo sapiens

   8-B fungi

   8-C protists

   8-D viruses

9. Many chemicals are solvents. One chemical is known as the *universal solvent*. Which chemical is it?

   9-A water

   9-B ammonia

   9-C bleach

   9-D sodium chloride

10. All cells have some sort of genetic material in them. In eukaryotic cells, where would the genetic material (DNA) be found?

    10-A ribosomes

    10-B mitochondria

    10-C liposome

    10-D nucleus

11. There is more than one type of nervous system in the human body. Which one is the brain a part of?

11-A peripheral nervous system

11-B central nervous system

11-C nerve core

11-D fascial nervous system

12. Plant cells and animal cells share some characteristics and differ in others. Which is the best example of a process in plants but not animals?

12-A making energy

12-B ATP synthesis

12-C photosynthesis

12-D replication

13. What is the branch of Earth science concerned with the atmosphere and weather?

13-A geometry

13-B atmospherology

13-C meteorology

13-D astrology

14. What are amino acids used to create?

14-A cytoplasm

14-B DNA

14-C cells

14-D proteins

15. Out of the planets, one has a "vertical" equator when compared to the others. Which one is it?

15-A Uranus

15-B Neptune

15-C Venus

15-D Jupiter

16. Which of the following is not in the Kingdom Plantae?

16-A Cactus

16-B Algae

16-C Oak Tree

16-D Sunflower

17. What is the primary difference between a cell membrane and a cell wall?

17-A A cell membrane is flexible, and a cell wall is rigid.

17-B A cell membrane is not found in plants, whereas a cell wall is.

17-C A cell membrane is not found in animals, whereas a cell wall is.

17-D A cell membrane is composed of protein, whereas a cell wall is composed of sugar.

18. Plants are autotrophs, meaning that they:

18-A Consume organic material produced by animals

18-B They produce their own food

18-C They are able to move by themselves

18-D They can automatically transform from a seed into a plant.

19. Which of the following is not true of a virus?

19-A Viruses have DNA.

19-B Viruses do not have a nucleus.

19-C Viruses cannot survive without water.

19-D Viruses can be infectious.

20. In the digestive system, the majority of nutrients are absorbed in the:

20-A esophagus

20-B stomach

20-C small Intestine

20-D large Intestine

GO ON

General Science Answer Key

1.	A	11.	B
2.	D	12.	C
3.	B	13.	C
4.	A	14.	D
5.	C	15.	A
6.	A	16.	B
7.	B	17.	A
8.	C	18.	B
9.	A	19.	C
10.	D	20.	C

Made in the USA
San Bernardino, CA
08 January 2019